MW01091779

C ALGORITHMS FOR REAL-TIME DSP

PAUL M. EMBREE

Prentice Hall PTR
Upper Saddle River, NJ 07458

Library of Congress Cataloging-in-Publication Data

Embree, Paul M.
C algorithms for real-time DSP / Paul M. Embree.
p. cm.
Includes bibliographical references.
ISBN 0-13-337353-3
1. C (Computer program language) 2. Computer algorithms. 3. Real-time data processing I. Title.
QA76.73.C15E63 1995
621.382'2'028552—dc20 95-4077
CIP

Acquisitions editor: Karen Gettman
Cover designer: Judith Leeds Design
Manufacturing buyer: Alexis R. Heydt
Compositor/Production services: Pine Tree Composition, Inc.

The publisher offers discounts on this book when ordered in bulk quantities.

For more information contact:
Corporate Sales Department
Prentice Hall PTR
One Lake Street
Upper Saddle River, New Jersey 07458

Phone: 800-382-3419
Fax: 201-236-7141
email: Corpsales@prenhall.com

Printed in the United States of America

10 9 8 7 6 5 4 3

ISBN: 0-13-337353-3

Prentice Hall International (UK) Limited, *London*
Prentice Hall of Australia Pty. Limited, *Sydney*
Prentice Hall Canada, Inc., *Toronto*
Prentice Hall Hispanoamericana, S.A., *Mexico*
Prentice Hall of India Private Limited, *New Delhi*
Prentice Hall of Japan, Inc., *Tokyo*
Simon & Schuster Asia Pte. Ltd., *Singapore*
Editora Prentice Hall do Brasil, Ltda., *Rio de Janeiro*

CONTENTS

PREFACE

Digital signal processing techniques have become the method of choice in signal processing as digital computers have increased in speed, convenience, and availability. As microprocessors have become less expensive and more powerful, the number of DSP applications which have become commonly available has exploded. Thus, some DSP microprocessors can now be considered commodity products. Perhaps the most visible high volume DSP applications are the so called "multimedia" applications in digital audio, speech processing, digital video, and digital communications. In many cases, these applications contain embedded digital signal processors where a host CPU works in a loosely coupled way with one or more DSPs to control the signal flow or DSP algorithm behavior at a real-time rate. Unfortunately, the development of signal processing algorithms for these specialized embedded DSPs is still difficult and often requires specialized training in a particular assembly language for the target DSP.

The tools for developing new DSP algorithms are slowly improving as the need to design new DSP applications more quickly becomes important. The C language is proving itself to be a valuable programming tool for real-time computationally intensive software tasks. C has high-level language capabilities (such as structures, arrays, and functions) as well as low-level assembly language capabilities (such as bit manipulation, direct hardware input/output, and macros) which makes C an ideal language for embedded DSP. Most of the manufacturers of digital signal processing devices (such as Texas Instruments, AT&T, Motorola, and Analog Devices) provide C compilers, simulators, and emulators for their parts. These C compilers offer standard C language with extensions for DSP to allow for very efficient code to be generated. For example, an inline assembly language capability is usually provided in order to optimize the performance of time critical parts of an application. Because the majority of the code is C, an application can be transferred to another processor much more easily than an all assembly language program.

This book is constructed in such a way that it will be most useful to the engineer who is familiar with DSP and the C language, but who is not necessarily an expert in both. All of the example programs in this book have been tested using standard C compilers in the UNIX and MS-DOS programming environments. In addition, the examples have been compiled utilizing the real-time programing tools of specific real-time embedded DSP microprocessors (Analog Devices' ADSP-21020 and ADSP-21062; Texas Instrument's TMS320C30 and TMS320C40; and AT&T's DSP32C) and then tested with real-time hardware using real world signals. All of the example programs presented in the text are provided in source code form on the IBM PC floppy disk included with the book.

The text is divided into several sections. Chapters 1 and 2 cover the basic principles of digital signal processing and C programming. Readers familiar with these topics may wish to skip one or both chapters. Chapter 3 introduces the basic real-time DSP programming techniques and typical programming environments which are used with DSP microprocessors. Chapter 4 covers the basic real-time filtering techniques which are the cornerstone of one-dimensional real-time digital signal processing. Finally, several real-time DSP applications are presented in Chapter 5, including speech compression, music signal processing, radar signal processing, and adaptive signal processing techniques.

The floppy disk included with this text contains C language source code for all of the DSP programs discussed in this book. The floppy disk has a high density format and was written by MS-DOS. The appendix and the READ.ME files on the floppy disk provide more information about how to compile and run the C programs. These programs have been tested using Borland's TURBO C (version 3 and greater) as well as Microsoft C (versions 6 and greater) for the IBM PC. Real-time DSP platforms using the Analog Devices ADSP-21020 and the ADSP-21062, the Texas Instruments TMS320C30, and the AT&T DSP32C have been used extensively to test the real-time performance of the algorithms.

ACKNOWLEDGMENTS

I thank the following people for their generous help: Laura Mercs for help in preparing the electronic manuscript and the software for the DSP32C; the engineers at Analog Devices (in particular Steve Cox, Marc Hoffman, and Hans Rempel) for their review of the manuscript as well as hardware and software support; Texas Instruments for hardware and software support; Jim Bridges at Communication Automation & Control, Inc., and Talal Itani at Domain Technologies, Inc.

Paul M. Embree

TRADEMARKS

CHAPTER 1

DIGITAL SIGNAL PROCESSING FUNDAMENTALS

Digital signal processing begins with a digital signal which appears to the computer as a sequence of digital values. Figure 1.1 shows an example of a digital signal processing operation or simple DSP system. There is an input sequence $x(n)$, the operator $\mathcal{O}\{\ \}$ and an output sequence, $y(n)$. A complete digital signal processing system may consist of many operations on the same sequence as well as operations on the result of operations. Because digital sequences are processed, all operators in DSP are discrete time operators (as opposed to continuous time operators employed by analog systems). Discrete time operators may be classified as *time-varying* or *time-invariant* and *linear* or *nonlinear*. Most of the operators described in this text will be time-invariant with the exception of adaptive filters which are discussed in Section 1.7. Linearity will be discussed in Section 1.2 and several nonlinear operators will be introduced in Section 1.5.

Operators are applied to sequences in order to effect the following results:

(1) Extract parameters or features from the sequence.
(2) Produce a similar sequence with particular features enhanced or eliminated.
(3) Restore the sequence to some earlier state.
(4) Encode or compress the sequence.

This chapter is divided into several sections. Section 1.1 deals with *sequences* of numbers: where and how they originate, their spectra, and their relation to continuous signals. Section 1.2 describes the common characteristics of *linear time-invariant operators* which are the most often used in DSP. Section 1.3 discusses the class of operators called *digital filters.* Section 1.4 introduces the *discrete Fourier transform* (DFTs and

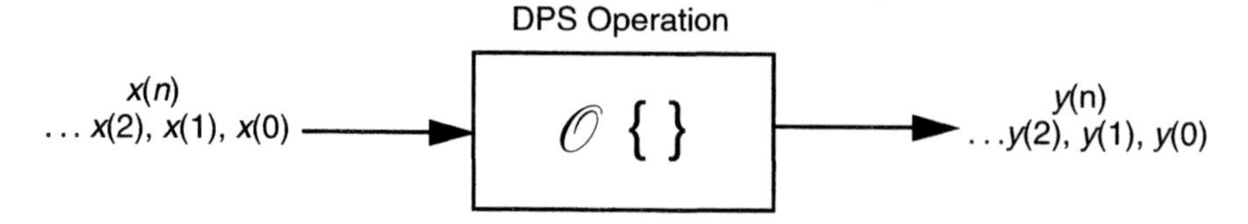

FIGURE 1.1 DSP operation.

FFTs). Section 1.5 describes the properties of commonly used *nonlinear operators.* Section 1.6 covers basic *probability theory* and *random processes* and discusses their application to signal processing. Finally, Section 1.7 discusses the subject of *adaptive digital filters.*

1.1 SEQUENCES

In order for the digital computer to manipulate a signal, the signal must have been sampled at some interval. Figure 1.2 shows an example of a continuous function of time which has been sampled at intervals of T seconds. The resulting set of numbers is called a *sequence.* If the continuous time function was $x(t)$, then the samples would be $x(nT)$ for n, an integer extending over some finite range of values. It is common practice to normalize the sample interval to 1 and drop it from the equations. The sequence then becomes $x(n)$. Care must be taken, however, when calculating power or energy from the sequences. The sample interval, including units of time, must be reinserted at the appropriate points in the power or energy calculations.

A sequence as a representation of a continuous time signal has the following important characteristics:

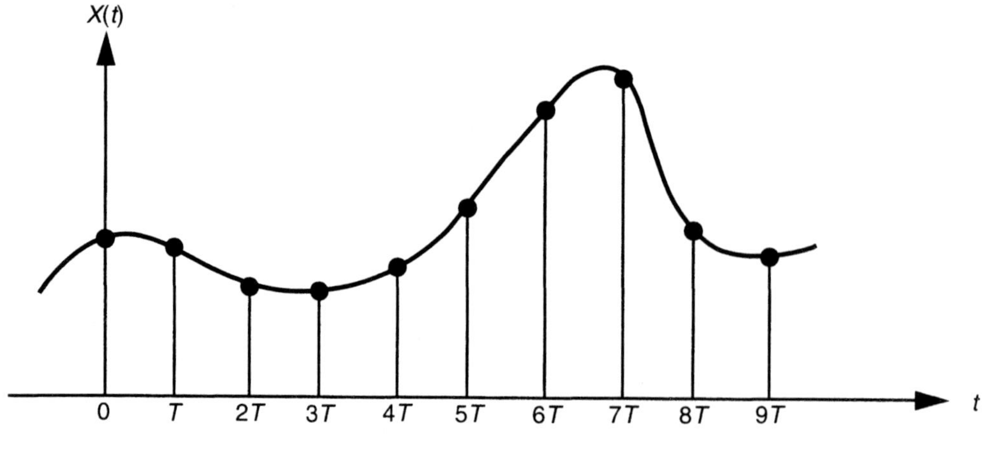

FIGURE 1.2 Sampling.

(1) The signal is sampled. It has finite value at only discrete points in time.

(2) The signal is truncated outside some finite length representing a finite time interval.

(3) The signal is quantized. It is limited to discrete steps in amplitude, where the step size and, therefore, the accuracy (or *signal fidelity*) depends on how many steps are available in the A/D converter and on the arithmetic precision (number of bits) of the digital signal processor or computer.

In order to understand the nature of the results that DSP operators produce, these characteristics must be taken into account. The effect of sampling will be considered in Section 1.1.1. Truncation will be considered in the section on the discrete Fourier transform (Section 1.4) and quantization will be discussed in Section 1.7.4.

1.1.1 The Sampling Function

The *sampling function* is the key to traveling between the continuous time and discrete time worlds. It is called by various names: the *Dirac delta function,* the *sifting function,* the *singularity function,* and the *sampling function* among them. It has the following properties:

$$\text{Property 1.} \int_{-\infty}^{\infty} f(t)\delta(t-\tau)dt = f(\tau). \tag{1.1}$$

$$\text{Property 2.} \int_{-\infty}^{\infty} \delta(t-\tau)dt = 1. \tag{1.2}$$

In the equations above, τ can be any real number.

To see how this function can be thought of as the ideal sampling function, first consider the realizable sampling function, $\Delta(t)$, illustrated in Figure 1.3. Its pulse width is one unit of time and its amplitude is one unit of amplitude. It clearly exhibits Property 2 of the sampling function. When $\Delta(t)$ is multiplied by the function to be sampled, however, the $\Delta(t)$ sampling function chooses not a single instant in time but a range from $-\frac{1}{2}$ to $+\frac{1}{2}$. As a result, Property 1 of the sampling function is not met. Instead the following integral would result:

$$\int_{-\infty}^{\infty} f(t)\Delta(t-\tau)dt = \int_{\tau-\frac{1}{2}}^{\tau+\frac{1}{2}} f(t)dt. \tag{1.3}$$

This can be thought of as a kind of smearing of the sampling process across a band which is related to the pulse width of $\Delta(t)$. A better approximation to the sampling function would be a function $\Delta(t)$ with a narrower pulse width. As the pulse width is narrowed, however, the amplitude must be increased. In the limit, the ideal sampling function must have infinitely narrow pulse width so that it samples at a single instant in time, and infinitely large amplitude so that the sampled signal still contains the same finite energy.

Figure 1.2 illustrates the sampling process at sample intervals of T. The resulting time waveform can be written

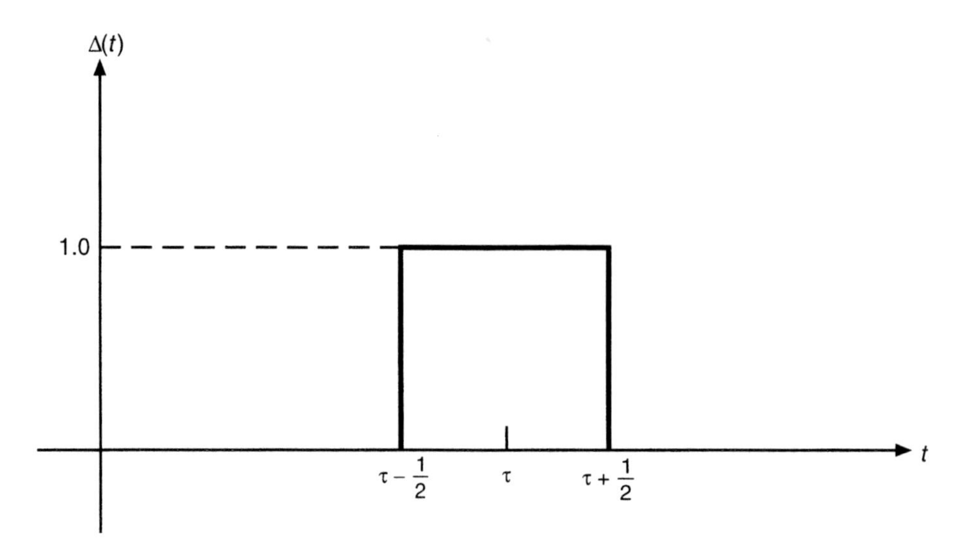

FIGURE 1.3 Realizable sampling function.

$$x_s(t) = \sum_{n=-\infty}^{\infty} x(t)\delta(t - nT). \tag{1.4}$$

The waveform that results from this process is impossible to visualize due to the infinite amplitude and zero width of the ideal sampling function. It may be easier to picture a somewhat less than ideal sampling function (one with very small width and very large amplitude) multiplying the continuous time waveform.

It should be emphasized that $x_s(t)$ is a continuous time waveform made from the superposition of an infinite set of continuous time signals $x(t)\delta(t - nT)$. It can also be written

$$x_s(t) = \sum_{n=-\infty}^{\infty} x(nT)\delta(t - nT) \tag{1.5}$$

since the sampling function gives a nonzero multiplier only at the values $t = nT$. In this last equation, the sequence $x(nT)$ makes its appearance. This is the set of numbers or *samples* on which almost all DSP is based.

1.1.2 Sampled Signal Spectra

Using *Fourier transform theory,* the frequency spectrum of the continuous time waveform $x(t)$ can be written

$$X(f) = \int_{-\infty}^{\infty} x(t)e^{-j2\pi ft}\,dt \tag{1.6}$$

and the time waveform can be expressed in terms of its spectrum as

$$x(t) = \int_{-\infty}^{\infty} X(f)e^{j2\pi ft}\,df. \tag{1.7}$$

Since this is true for any continuous function of time, $x(t)$, it is also true for $x_s(t)$.

$$X_s(f) = \int_{-\infty}^{\infty} x_s(t)e^{-j2\pi ft}\,dt. \tag{1.8}$$

Replacing $x_s(t)$ by the sampling representation

$$X_s(f) = \int_{-\infty}^{\infty}\left[\sum_{n=-\infty}^{\infty} x(t)\delta(t-nT)\right]e^{-j2\pi ft}\,dt. \tag{1.9}$$

The order of the summation and integration can be interchanged and Property 1 of the sampling function applied to give

$$X_s(f) = \sum_{n=-\infty}^{\infty} x(nT)e^{-j2\pi fnT}. \tag{1.10}$$

This equation is the exact form of a Fourier series representation of $X_s(f)$, a periodic function of frequency having period $1/T$. The coefficients of the Fourier series are $x(nT)$ and they can be calculated from the following integral:

$$x(nT) = T\int_{-\frac{1}{2T}}^{\frac{1}{2T}} X_2(f)e^{j2\pi fnT}\,df. \tag{1.11}$$

The last two equations are a Fourier series pair which allow calculation of either the time signal or frequency spectrum in terms of the opposite member of the pair. Notice that the use of the problematic signal $x_s(t)$ is eliminated and the sequence $x(nT)$ can be used instead.

1.1.3 Spectra of Continuous Time and Discrete Time Signals

By evaluating Equation (1.7) at $t = nT$ and setting the result equal to the right-hand side of Equation (1.11) the following relationship between the two spectra is obtained:

$$x(nT) = \int_{-\infty}^{\infty} X(f)e^{j2\pi fnT}\,df = T\int_{-\frac{1}{2T}}^{\frac{1}{2T}} X_s(f)e^{j2\pi fnT}\,df. \tag{1.12}$$

The right-hand side of Equation (1.7) can be expressed as the infinite sum of a set of integrals with finite limits

$$x(nT) = \sum_{m=-\infty}^{\infty} T\int_{\frac{2m-1}{2T}}^{\frac{2m+1}{2T}} X(f)e^{j2\pi fnT}\,df. \tag{1.13}$$

By changing variables to $\lambda = f - m/T$ (substituting $f = \lambda + m/T$ and $df = d\lambda$)

$$x(nT) = \sum_{m=-\infty}^{\infty} \int_{-\frac{1}{2T}}^{\frac{1}{2T}} X(\lambda + \frac{m}{T}) e^{j2\pi\lambda nT} e^{j2\pi\frac{m}{T}nT} d\lambda. \tag{1.14}$$

Moving the summation inside the integral, recognizing that $e^{j2\pi mn}$ (for all integers m and n) is equal to 1, and equating everything inside the integral to the similar part of Equation (1.11) give the following relation:

$$X_s(f) = \sum_{m=-\infty}^{\infty} X(f + \frac{m}{T}). \tag{1.15}$$

Equation (1.15) shows that the sampled time frequency spectrum is equal to an infinite sum of shifted replicas of the continuous time frequency spectrum overlaid on each other. The shift of the replicas is equal to the sample frequency, $1/T$. It is interesting to examine the conditions under which the two spectra are equal to each other, at least for a limited range of frequencies. In the case where there are no spectral components of frequency greater than $1/2T$ in the original continuous time waveform, the two spectra

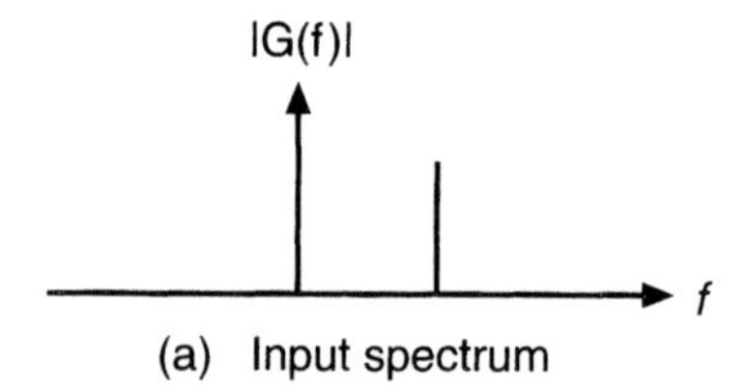

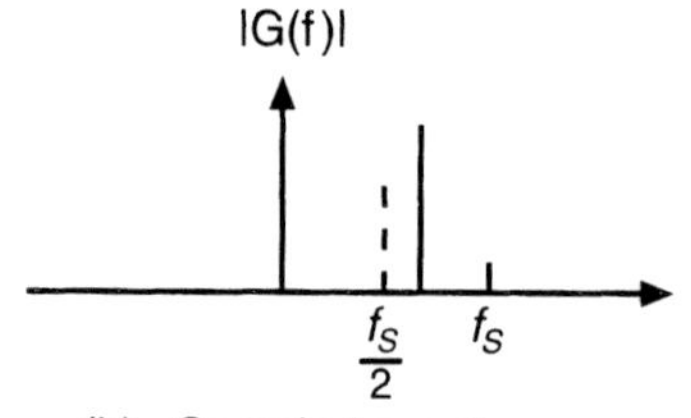

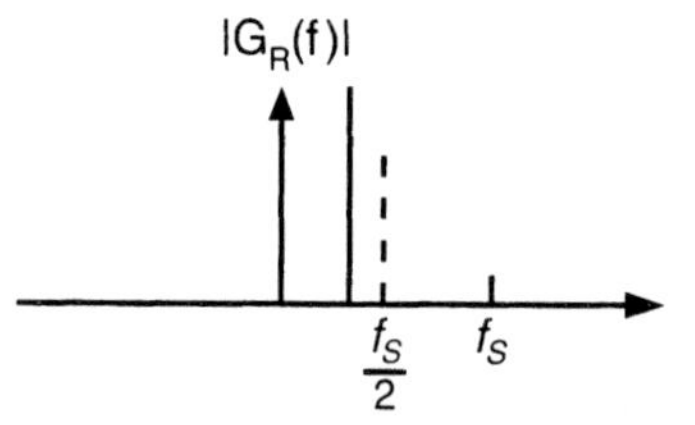

FIGURE 1.4 Aliasing in the frequency domain. **(a)** Input spectrum. **(b)** Sampled spectrum. **(c)** Reconstructed spectrum.

are equal over the frequency range $f = -\frac{1}{2T}$ to $f = +\frac{1}{2T}$. Of course, the sampled time spectrum will repeat this same set of amplitudes periodically for all frequencies, while the continuous time spectrum is identically zero for all frequencies outside the specified range.

The *Nyquist sampling criterion* is based on the derivation just presented and asserts that a continuous time waveform, when sampled at a frequency greater than twice the maximum frequency component in its spectrum, can be reconstructed completely from the sampled waveform. Conversely, if a continuous time waveform is sampled at a frequency lower than twice its maximum frequency component a phenomenon called *aliasing* occurs. If a continuous time signal is reconstructed from an aliased representation, distortions will be introduced into the result and the degree of distortion is dependent on the degree of aliasing. Figure 1.4 shows the spectra of sampled signals without aliasing and with aliasing. Figure 1.5 shows the reconstructed waveforms of an aliased signal.

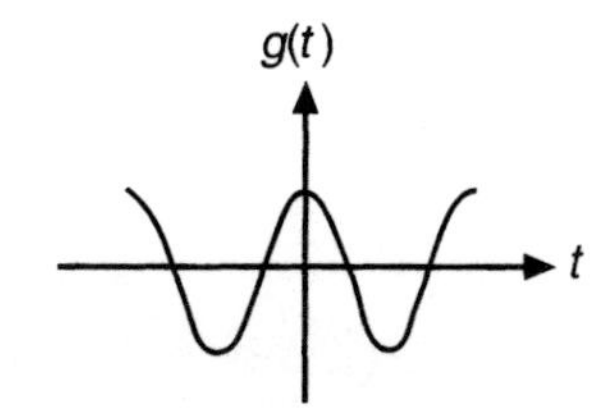

(a) Input continuous time signal

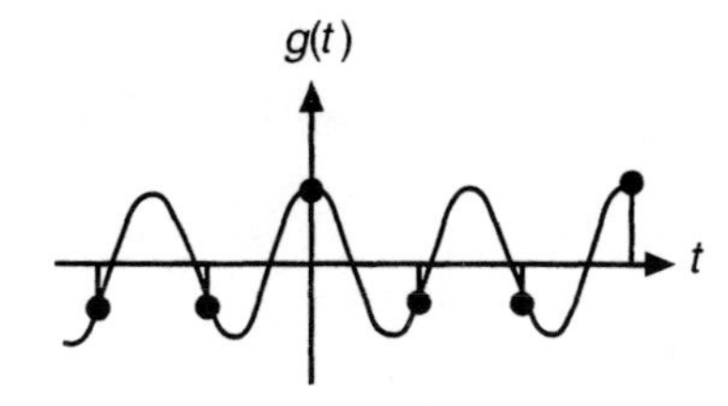

(b) Sampled signal

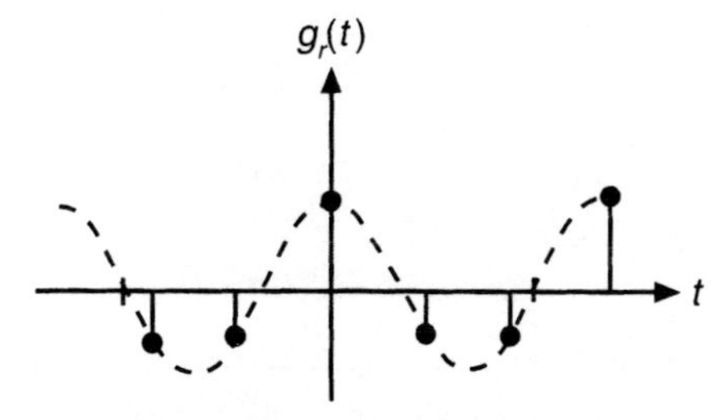

(c) Reconstructed signal

FIGURE 1.5 Aliasing in the time domain. **(a)** Input continuous time signal. **(b)** Sampled signal. **(c)** Reconstructed signal.

1.2 LINEAR TIME-INVARIANT OPERATORS

The most commonly used DSP operators are *linear* and *time-invariant* (or LTI). The linearity property is stated as follows:

Given $x(n)$, a finite sequence, and $\mathbb{O}\{\ \}$, an operator in n-space, let

$$y(n) = \mathbb{O}\{x(n)\}. \tag{1.16}$$

If

$$x(n) = ax_1(n) + bx_2(n) \tag{1.17}$$

where a and b are constant with respect to n, then, if $\mathbb{O}\{\ \}$ is a linear operator

$$y(n) = a\mathbb{O}\{x_1(n)\} + b\mathbb{O}\{x_2(n)\}. \tag{1.18}$$

The time-invariant property means that if

$$y(n) = \mathbb{O}\{x(n)\}$$

then the shifted version gives the same response or

$$y(n-m) = \mathbb{O}\{x(n-m)\}. \tag{1.19}$$

Another way to state this property is that if $x(n)$ is periodic with period N such that

$$x(n+N) = x(n)$$

then if $\mathbb{O}\{\ \}$ is a time-invariant operator in n space

$$\mathbb{O}\{x(n+N)\} = \mathbb{O}x(n)\}.$$

Next, the LTI properties of the operator $\mathbb{O}\{\ \}$ will be used to derive an expression and method of calculation for $\mathbb{O}\{x(n)\}$. First, the impulse sequence can be used to represent $x(n)$ in a different manner,

$$x(n) = \sum_{m=-\infty}^{\infty} x(m)u_0(n-m). \tag{1.20}$$

This is because

$$u_0(n-m) = \begin{cases} 1, & n = m \\ 0, & \text{otherwise.} \end{cases} \tag{1.21}$$

The impulse sequence acts as a sampling or sifting function on the function $x(m)$, using the dummy variable m to sift through and find the single desired value $x(n)$. Now this somewhat devious representation of $x(n)$ is substituted into the operator Equation (1.16):

$$y(n) = \mathcal{O}\left\{ \sum_{m=-\infty}^{\infty} x(m)u_0(n-m) \right\}. \tag{1.22}$$

Recalling that $\mathcal{O}\{\ \}$ operates only on functions of n and using the linearity property

$$y(n) = \sum_{m=-\infty}^{\infty} x(m)\mathcal{O}\{u_0(n-m)\}. \tag{1.23}$$

Every operator has a set of outputs that are its response when an impulse sequence is applied to its input. The impulse response is represented by $h(n)$ so that

$$h(n) = \mathcal{O}\{u_0(n)\}. \tag{1.24}$$

This impulse response is a sequence that has special significance for $\mathcal{O}\{\ \}$, since it is the sequence that occurs at the output of the block labeled $\mathcal{O}\{\ \}$ in Figure 1.1 when an impulse sequence is applied at the input. By time invariance it must be true that

$$h(n-m) = \mathcal{O}\{u_0(n-m)\} \tag{1.25}$$

so that

$$y(n) = \sum_{m=-\infty}^{\infty} x(m)\,h(n-m). \tag{1.26}$$

Equation (1.26) states that $y(n)$ is equal to the convolution of $x(n)$ with the impulse response $h(n)$. By substituting $m = n - p$ into Equation (1.26) an equivalent form is derived

$$y(n) = \sum_{p=-\infty}^{\infty} h(p)\,x(n-p). \tag{1.27}$$

It must be remembered that m and p are dummy variables and are used for purposes of the summation only. From the equations just derived it is clear that the impulse response completely characterizes the operator $\mathcal{O}\{\ \}$ and can be used to label the block representing the operator as in Figure 1.6.

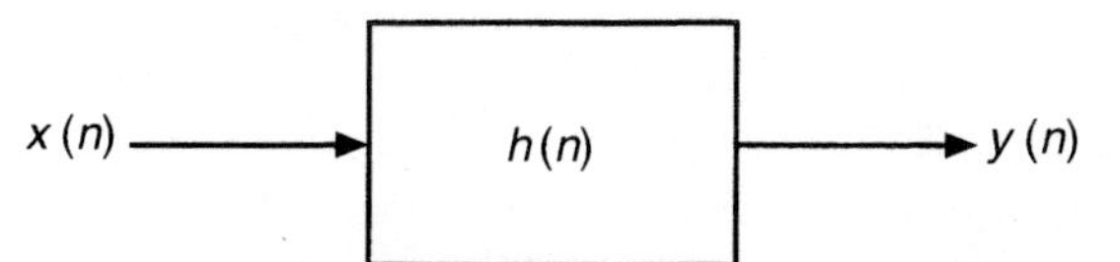

FIGURE 1.6 Impulse response representation of an operator.

1.2.1 Causality

In the mathematical descriptions of sequences and operators thus far, it was assumed that the impulse responses of operators may include values that occur before any applied input stimulus. This is the most general form of the equations and has been suitable for the development of the theory to this point. However, it is clear that no physical system can produce an output in response to an input that has not yet been applied. Since DSP operators and sequences have their basis in physical systems, it is more useful to consider that subset of operators and sequences that can exist in the real world.

The first step in representing realizable sequences is to acknowledge that any sequence must have started at some time. Thus, it is assumed that any element of a sequence in a realizable system whose time index is less than zero has a value of zero. Sequences which start at times later than this can still be represented, since an arbitrary number of their beginning values can also be zero. However, the earliest true value of any sequence must be at a value of n that is greater than or equal to zero. This attribute of sequences and operators is called *causality,* since it allows all attributes of the sequence to be caused by some physical phenomenon. Clearly, a sequence that has already existed for infinite time lacks a cause, as the term is generally defined.

Thus, the convolution relation for causal operators becomes:

$$y(n) = \sum_{m=0}^{\infty} h(m)\, x(n-m). \tag{1.28}$$

This form follows naturally since the impulse response is a sequence and can have no values for m less than zero.

1.2.2 Difference Equations

All discrete time, linear, causal, time-invariant operators can be described in theory by the Nth order difference equation

$$\sum_{m=0}^{N-1} a_m y(n-m) = \sum_{p=0}^{N-1} b_p x(n-p) \tag{1.29}$$

where $x(n)$ is the stimulus for the operator and $y(n)$ is the results or output of the operator. The equation remains completely general if all coefficients are normalized by the value of a_0 giving

$$y(n) + \sum_{m=1}^{N-1} a_m y(n-m) = \sum_{p=0}^{N-1} b_p x(n-p) \tag{1.30}$$

and the equivalent form

$$y(n) = \sum_{p=0}^{N-1} b_p x(n-p) - \sum_{m=1}^{N-1} a_m y(n-m) \tag{1.31}$$

or

$$\begin{aligned} y(n) &= b_0x(n) + b_1x(n-1) + b_2x(n-2)\ldots \\ &+ b_{N-1}x(n-N+1) - a_1y(n-1) - a_2y(n-2) \\ &- \ldots - a_{N-1}y(n-N+1). \end{aligned} \tag{1.32}$$

To represent an operator properly may require a very high value of N, and for some complex operators N may have to be infinite. In practice, the value of N is kept within limits manageable by a computer; there are often approximations made of a particular operator to make N an acceptable size.

In Equations (1.30) and (1.31) the terms $y(n - m)$ and $x(n - p)$ are shifted or delayed versions of the functions $y(n)$ and $x(n)$, respectively. For instance, Figure 1.7 shows a sequence $x(n)$ and $x(n - 3)$, which is the same sequence delayed by three sample periods. Using this delaying property and Equation (1.32), a structure or flow graph can be constructed for the general form of a discrete time LTI operator. This structure is shown in Figure 1.8. Each of the boxes is a delay element with unity gain. The coefficients are shown next to the legs of the flow graph to which they apply. The circles enclosing the summation symbol (Σ) are adder elements.

1.2.3 The *z*-Transform Description of Linear Operators

There is a linear transform—called the *z-transform*—which is as useful to discrete time analysis as the Laplace transform is to continuous time analysis. Its definition is

$$\mathcal{Z}\{x(n)\} = \sum_{n=0}^{\infty} x(n)z^{-n} \tag{1.33}$$

where the symbol $\mathcal{Z}\{\ \}$ stands for "z-transform of," and the z in the equation is a complex number. One of the most important properties of the z-transform is its relationship to time

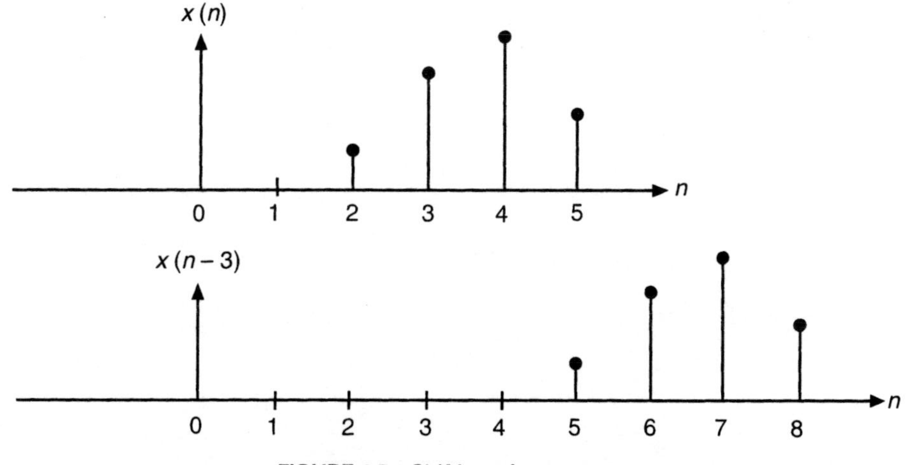

FIGURE 1.7 Shifting of a sequence.

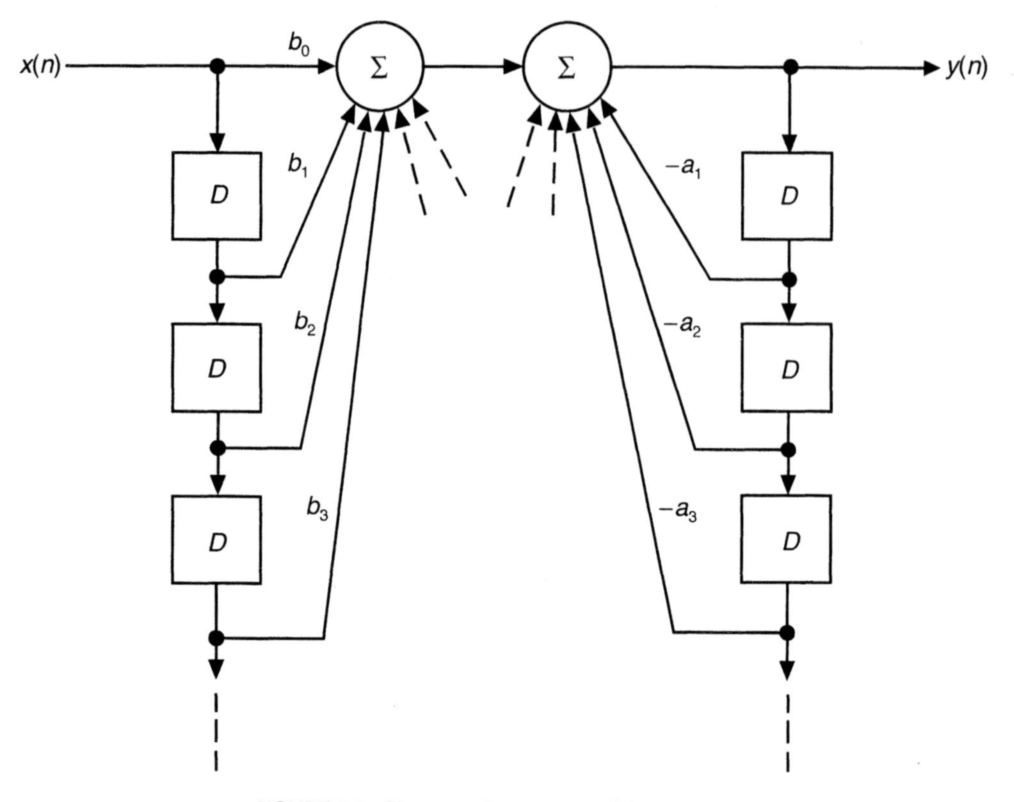

FIGURE 1.8 Flow graph structure of linear operators.

delay in sequences. To show this property take a sequence, $x(n)$, with a z-transform as follows:

$$\mathcal{Z}\{x(n)\} = X(z) = \sum_{n=0}^{\infty} x(n)z^{-n}. \tag{1.34}$$

A shifted version of this sequence has a z-transform:

$$\mathcal{Z}\{x(n-p)\} = \sum_{n=0}^{\infty} x(n-p)z^{-n}. \tag{1.35}$$

By letting $m = n - p$ substitution gives:

$$\mathcal{Z}\{x(n-p)\} = \sum_{m=0}^{\infty} x(m)z^{-(m+p)} \tag{1.36}$$

$$= z^{-p} \sum_{m=0}^{\infty} x(m) z^{-m}. \tag{1.37}$$

But comparing the summation in this last equation to Equation (1.33) for the z-transform of $x(n)$, it can be seen that

$$\mathcal{Z}\{x(n-p)\} = z^{-p}\mathcal{Z}\{x(n)\} = z^{-p}X(z). \tag{1.38}$$

This property of the z-transform can be applied to the general equation for LTI operators as follows:

$$\mathcal{Z}\left\{ y(n) + \sum_{p=1}^{\infty} a_p y(n-p) \right\} = z^{-p}\, \mathcal{Z}\left\{ \sum_{q=0}^{\infty} b_q x(n-q) \right\}. \tag{1.39}$$

Since the z-transform is a linear transform, it possesses the distributive and associative properties. Equation (1.39) can be simplified as follows:

$$\mathcal{Z}\{y(n)\} + \sum_{p=1}^{\infty} a_p \mathcal{Z}\{y(n-p)\} = \sum_{q=0}^{\infty} b_q \mathcal{Z}\{x(n-p)\}. \tag{1.40}$$

Using the shift property of the z-transform (Equation (1.38))

$$Y(z) + \sum_{p=1}^{\infty} a_p z^{-p} Y(z) = \sum_{q=0}^{\infty} b_q z^{-q} X(z) \tag{1.41}$$

$$Y(z)\left[1 + \sum_{p=1}^{\infty} a_p z^{-p} \right] = X(z)\left[\sum_{q=0}^{\infty} b_q z^{-q} \right]. \tag{1.42}$$

Finally, Equation (1.42) can be rearranged to give the transfer function in the z-transform domain:

$$H(z) = \frac{Y(z)}{X(z)} = \frac{\sum_{q=0}^{\infty} b_q z^{-q}}{1 + \sum_{p=1}^{\infty} a_p z^{-p}}. \tag{1.43}$$

Using Equation (1.41), Figure 1.8 can be redrawn in the z-transform domain and this structure is shown in Figure 1.9. The flow graphs are identical if it is understood that a multiplication by z^{-1} in the transform domain is equivalent to a delay of one sampling time interval in the time domain.

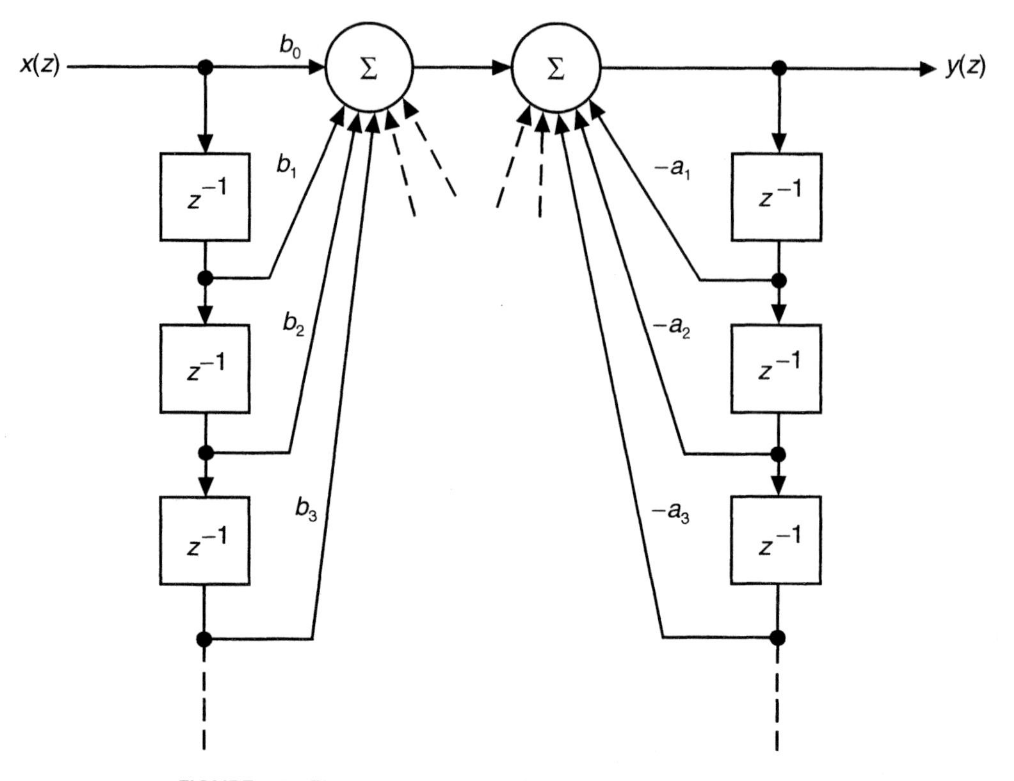

FIGURE 1.9 Flow graph structure for the *z*-transform of an operator.

1.2.4 Frequency Domain Transfer Function of an Operator

Taking the Fourier transform of both sides of Equation (1.28) (which describes any LTI causal operator) results in the following:

$$\mathcal{F}\{y(n)\} = \sum_{m=0}^{\infty} h(m)\mathcal{F}\{x(n-m)\}. \tag{1.44}$$

Using one of the properties of the Fourier transform

$$\mathcal{F}\{x(n-m)\} = e^{-j2\pi fm}\mathcal{F}\{x(n)\}. \tag{1.45}$$

From Equation (1.45) it follows that

$$Y(f) = \sum_{m=0}^{\infty} h(m)e^{-j2\pi fm}X(f), \tag{1.46}$$

or dividing both sides by $X(f)$

$$\frac{Y(f)}{X(f)} = \sum_{m=0}^{\infty} h(m)e^{-j2\pi fm}, \tag{1.47}$$

which is easily recognized as the Fourier transform of the series $h(m)$. Rewriting this equation

$$\frac{Y(f)}{X(f)} = H(f) = \mathcal{F}\{h(m)\}. \tag{1.48}$$

Figure 1.10 shows the time domain block diagram of Equation (1.48) and Figure 1.11 shows the Fourier transform (or frequency domain) block diagram and equation. The frequency domain description of a linear operator is often used to describe the operator. Most often it is shown as an amplitude and a phase angle plot as a function of the variable f (sometimes normalized with respect to the sampling rate, $1/T$).

1.2.5 Frequency Response from the *z*-Transform Description

Recall the Fourier transform pair

$$X_s(f) = \sum_{n=-\infty}^{\infty} x(nT)e^{-j2\pi fnT} \tag{1.49}$$

and

$$x(nT) = \int_{-\infty}^{\infty} X_s(f)e^{j2\pi fnT}\,df. \tag{1.50}$$

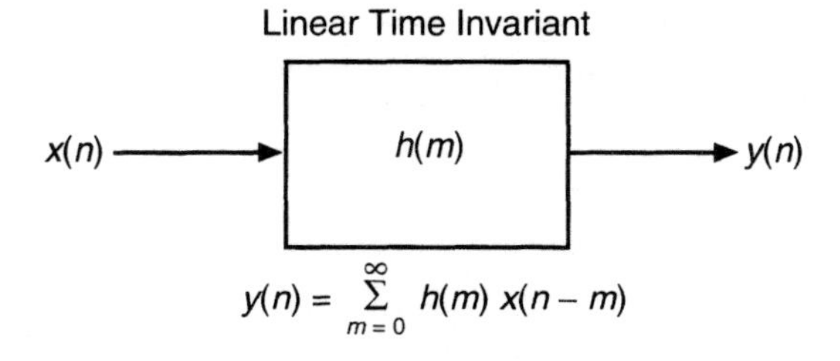

FIGURE 1.10 Time domain block diagram of LTI system.

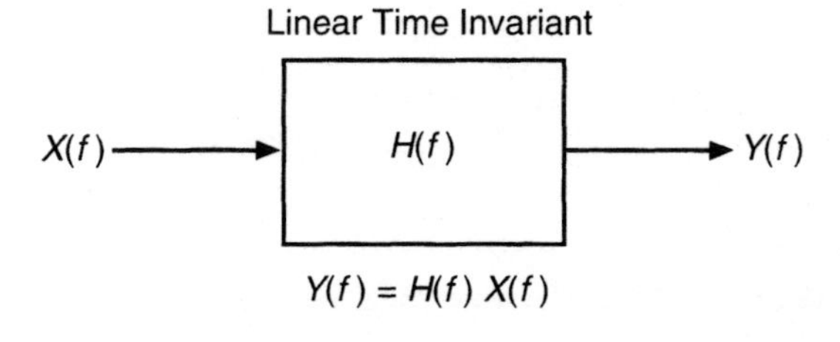

FIGURE 1.11 Frequency block diagram of LTI system.

In order to simplify the notation, the value of T, the period of the sampling waveform, is normalized to be equal to one.

Now compare Equation (1.49) to the equation for the z-transform of $x(n)$ as follows:

$$X(z) = \sum_{n=0}^{\infty} x(n) z^{-n}. \tag{1.51}$$

Equations (1.49) and (1.51) are equal for sequences $x(n)$ which are causal (i.e., $x(n) = 0$ for all $n < 0$) if z is set as follows:

$$z = e^{j2\pi f}. \tag{1.52}$$

A plot of the locus of values for z in the complex plane described by Equation (1.52) is shown in Figure 1.12. The plot is a circle of unit radius. Thus, the z-transform of a causal sequence, $x(n)$, when evaluated on the unit circle in the complex plane, is equivalent to the frequency domain representation of the sequence. This is one of the properties of the z-transform which make it very useful for discrete signal analysis.

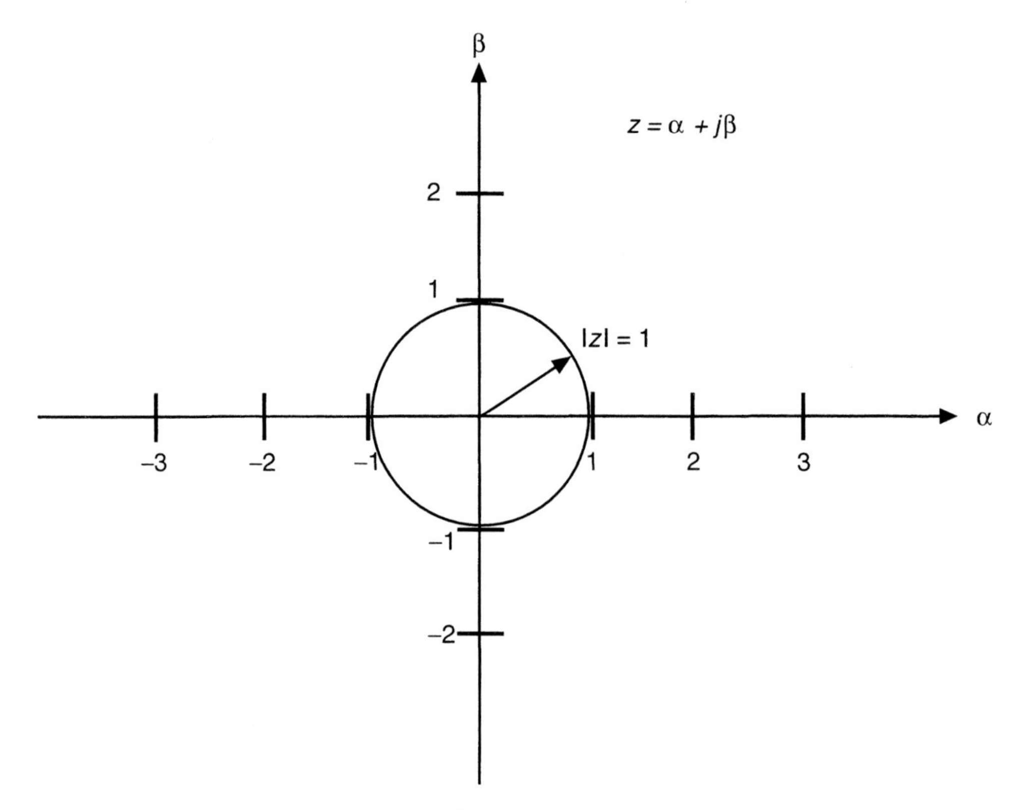

FIGURE 1.12 The unit circle in the z-plane.

Summarizing the last few paragraphs, the impulse response of an operator is simply a sequence, $h(m)$, and the Fourier transform of this sequence is the frequency response of the operator. The z-transform of the sequence $h(m)$, called $H(z)$, can be evaluated on the unit circle to yield the frequency domain representation of the sequence. This can be written as follows:

$$H(z)\Big|_{z=e^{j2\pi f}} = H(f). \tag{1.53}$$

1.3 DIGITAL FILTERS

The linear operators that have been presented and analyzed in the previous sections can be thought of as *digital filters*. The concept of *filtering* is an analogy between the action of a physical strainer or sifter and the action of a linear operator on sequences when the operator is viewed in the frequency domain. Such a filter might allow certain frequency components of the input to pass unchanged to the output while blocking other components. Naturally, any such action will have its corresponding result in the time domain. This view of linear operators opens a wide area of theoretical analysis and provides increased understanding of the action of digital systems.

There are two broad classes of digital filters. Recall the difference equation for a general operator:

$$y(n) = \sum_{q=0}^{Q-1} b_q x(n-q) - \sum_{p=1}^{P-1} a_p y(n-p). \tag{1.54}$$

Notice that the infinite sums have been replaced with finite sums. This is necessary in order that the filters can be physically realizable.

The first class of digital filters have a_p equal to 0 for all p. The common name for filters of this type is *finite impulse response* (FIR) filters, since their response to an impulse dies away in a finite number of samples. These filters are also called *moving average* (or MA) filters, since the output is simply a weighted average of the input values.

$$y(n) = \sum_{q=0}^{Q-1} b_q x(n-q). \tag{1.55}$$

There is a window of these weights (b_q) that takes exactly the Q most recent values of $x(n)$ and combines them to produce the output.

The second class of digital filters are *infinite impulse response* (IIR) filters. This class includes both *autoregressive* (AR) filters and the most general form, autoregressive moving average (*ARMA*) filters. In the AR case all b_q for $q = 1$ to $Q - 1$ are set to 0.

$$y(n) = x(n) - \sum_{p=1}^{P-1} a_p y(n-p) \tag{1.56}$$

For ARMA filters, the more general Equation (1.54) applies. In either type of IIR filter, a single-impulse response at the input can continue to provide output of infinite duration with a given set of coefficients. Stability can be a problem for IIR filters, since with poorly chosen coefficients, the output can grow without bound for some inputs.

1.3.1 Finite Impulse Response (FIR) Filters

Restating the general equation for FIR filters

$$y(n)=\sum_{q=0}^{Q-1} b_q x(n-q). \tag{1.57}$$

Comparing this equation with the convolution relation for linear operators

$$y(n)=\sum_{m=0}^{\infty} h(m)\,x(n-m),$$

one can see that the coefficients in an FIR filter are identical to the elements in the impulse response sequence if this impulse response is finite in length.

$$b_q = h(q) \qquad \text{for } q=0,1,2,3,\ldots,Q-1.$$

This means that if one is given the impulse response sequence for a linear operator with a finite impulse response one can immediately write down the FIR filter coefficients. However, as was mentioned at the start of this section, filter theory looks at linear operators primarily from the frequency domain point of view. Therefore, one is most often given the desired frequency domain response and asked to determine the FIR filter coefficients.

There are a number of methods for determining the coefficients for FIR filters given the frequency domain response. The two most popular FIR filter design methods are listed and described briefly below.

1. *Use of the DFT on the sampled frequency response.* In this method the required frequency response of the filter is sampled at a frequency interval of $1/T$ where T is the time between samples in the DSP system. The inverse discrete Fourier transform (see section 1.4) is then applied to this sampled response to produce the impulse response of the filter. Best results are usually achieved if a smoothing window is applied to the frequency response before the inverse DFT is performed. A simple method to obtain FIR filter coefficients based on the Kaiser window is described in section 4.1.2 in chapter 4.

2. *Optimal mini-max approximation using linear programming techniques.* There is a well-known program written by Parks and McClellan (1973) that uses the REMEZ exchange algorithm to produce an optimal set of FIR filter coefficients, given the required frequency response of the filter. The Parks-McClellan program is available on the IEEE digital signal processing tape or as part of many of the filter design packages available for personal computers. The program is also printed in several DSP texts (see Elliot 1987 or

Rabiner and Gold 1975). The program REMEZ.C is a C language implementation of the Parks-McClellan program and is included on the enclosed disk. An example of a filter designed using the REMEZ program is shown at the end of section 4.1.2 in chapter 4.

The design of digital filters will not be considered in detail here. Interested readers may wish to consult references listed at the end of this chapter giving complete descriptions of all the popular techniques.

The frequency response of FIR filters can be investigated by using the transfer function developed for a general linear operator:

$$H(z)=\frac{Y(z)}{X(z)}=\frac{\sum_{q=0}^{Q-1} b_q z^{-q}}{1+\sum_{p=1}^{P-1} a_p z^{-p}}. \tag{1.58}$$

Notice that the sums have been made finite to make the filter realizable. Since for FIR filters the a_p are all equal to 0, the equation becomes:

$$H(z)=\frac{Y(z)}{X(z)}=\sum_{q=0}^{Q-1} b_q z^{-q}. \tag{1.59}$$

The Fourier transform or frequency response of the transfer function is obtained by letting $z=e^{j2\pi f}$, which gives

$$H(f)=H(z)\big|_{z=e^{j2\pi f}}=\sum_{q=0}^{Q-1} b_q e^{-j2\pi fq}. \tag{1.60}$$

This is a polynomial in powers of z^{-1} or a sum of products of the form

$$H(z)=b_0+b_1z^{-1}+b_2z^{-2}+b_3z^{-3}+\ldots+b_{Q-1}z^{-(Q-1)}.$$

There is an important class of FIR filters for which this polynomial can be factored into a product of sums from

$$H(z)=\prod_{m=0}^{M-1}(z^{-2}+\alpha_m z^{-1}+\beta_m)\prod_{n=0}^{N-1}(z^{-1}+\gamma_n). \tag{1.61}$$

This expression for the transfer function makes explicit the values of the variable z^{-1} which cause $H(z)$ to become zero. These points are simply the roots of the quadratic equation

$$0=z^{-2}+\alpha_m z^{-1}+\beta_m,$$

which in general provides complex conjugate zero pairs, and the values γ_n which provide single zeros.

In many communication and image processing applications it is essential to have filters whose transfer functions exhibit a phase characteristic that changes linearly with a change in frequency. This characteristic is important because it is the phase transfer relationship that gives minimum distortion to a signal passing through the filter. A very useful feature of FIR filters is that for a simple relationship of the coefficients, b_q, the resulting filter is guaranteed to have a linear phase response. The derivation of the relationship which provides a linear phase filter follows.

A linear phase relationship to frequency means that

$$H(f) = |H(f)|\, e^{j[\alpha f + \beta]},$$

where α and β are constants. If the transfer function of a filter can be separated into a real function of f multiplied by a phase factor $e^{j[\alpha f + \beta]}$, then this transfer function will exhibit linear phase.

Taking the FIR filter transfer function:

$$H(z) = b_0 + b_1 z^{-1} + b_2 z^{-2} + b_3 z^{-3} + \ldots + b_{Q-1} z^{-(Q-1)}$$

and replacing z by $e^{j2\pi f}$ to give the frequency response

$$H(f) = b_0 + b_1 e^{-j2\pi f} + b_2 e^{-j2\pi(2f)} + \ldots + b_{Q-1} e^{-j2\pi(Q-1)f}.$$

Factoring out the factor $e^{-j2\pi(Q-1)f/2}$ and letting ζ equal $(Q-1)/2$ gives

$$H(f) = e^{-j2\pi\zeta f}\left\{b_0 e^{j2\pi\zeta f} + b_1 e^{j2\pi(\zeta-1)f} + b_2 e^{j2\pi(\zeta-2)f} + \ldots + b_{Q-2} e^{-j2\pi(\zeta-1)f} + b_{Q-1} e^{-j2\pi\zeta f}\right\}.$$

Combining the coefficients with complex conjugate phases and placing them together in brackets

$$\begin{aligned} H(f) = e^{-j2\pi\zeta f}\Big\{ & \left[b_0 e^{j2\pi\zeta} + b_{Q-1} e^{-j2\pi\zeta f}\right] \\ & + \left[b_1 e^{j2\pi(\zeta-1)f} + b_{Q-2} e^{-j2\pi(\zeta-1)f}\right] \\ & + \left[b_2 e^{j2\pi(\zeta-2)f} + b_{Q-3} e^{-j2\pi(\zeta-2)f}\right] \\ & + \ldots \Big\} \end{aligned}$$

If each pair of coefficients inside the brackets is set equal as follows:

$$\begin{aligned} b_0 &= b_{Q-1} \\ b_1 &= b_{Q-2} \\ b_2 &= b_{Q-3}, \text{ etc.} \end{aligned}$$

Each term in brackets becomes a cosine function and the linear phase relationship is achieved. This is a common characteristic of FIR filter coefficients.

1.3.2 Infinite Impulse Response (IIR) Filters

Repeating the general equation for IIR filters

$$y(n) = \sum_{q=0}^{Q-1} b_q x(n-q) - \sum_{p=1}^{P-1} a_p y(n-p).$$

The z-transform of the transfer function of an IIR filter is

$$H(z) = \frac{Y(z)}{X(z)} = \frac{\sum_{q=0}^{Q-1} b_q z^{-q}}{1 + \sum_{p=1}^{P-1} a_p z^{-p}}.$$

No simple relationship exists between the coefficients of the IIR filter and the impulse response sequence such as that which exists in the FIR case. Also, obtaining linear phase IIR filters is not a straightforward coefficient relationship as is the case for FIR filters. However, IIR filters have an important advantage over FIR structures: In general, IIR filters require fewer coefficients to approximate a given filter frequency response than do FIR filters. This means that results can be computed faster on a general purpose computer or with less hardware in a special purpose design. In other words, IIR filters are computationally efficient. The disadvantage of the recursive realization is that IIR filters are much more difficult to design and implement. Stability, roundoff noise, and sometimes phase nonlinearity must be considered carefully in all but the most trivial IIR filter designs.

The direct form IIR filter realization shown in Figure 1.9, though simple in appearance, can have severe response sensitivity problems because of coefficient quantization, especially as the order of the filter increases. To reduce these effects, the transfer function is usually decomposed into second order sections and then realized as cascade sections. The C language implementation given in section 4.1.3 uses single precision floating-point numbers in order to avoid coefficient quantization effects associated with fixed-point implementations that can cause instability and significant changes in the transfer function.

IIR digital filters can be designed in many ways, but by far the most common IIR design method is the *bilinear transform.* This method relies on the existence of a known s-domain transfer function (or Laplace transform) of the filter to be designed. The s-domain filter coefficients are transformed into equivalent z-domain coefficients for use in an IIR digital filter. This might seem like a problem, since s-domain transfer functions are just as hard to determine as z-domain transfer functions. Fortunately, Laplace transform methods and s-domain transfer functions were developed many years ago for designing analog filters as well as for modeling mechanical and even biological systems. Thus, many tables of s-domain filter coefficients are available for almost any type of filter function (see the references for a few examples). Also, computer programs are available to generate coefficients for many of the common filter types (see the books by Jong, Anoutino, Stearns (1993), Embree (1991), or one of the many filter design packages available for personal computers). Because of the vast array of available filter tables, the large number of filter types, and because the design and selection of a filter requires careful examination of all the requirements (passband ripple, stopband attenuation as well as phase response in some cases), the subject of s-domain IIR filter design will not be covered in this book. However, several IIR filter designs with exact z-domain coefficients are given in the examples in section 4.1 and on the enclosed disk.

1.3.3 Examples of Filter Responses

As an example of the frequency response of an FIR filter with very simple coefficients, take the following moving average difference equation:

$$y(n) = 0.11\,x(n) + 0.22\,x(n-1) + 0.34\,x(n-2) \\ + 0.22\,x(n-3) + 0.11\,x(n-4).$$

One would suspect that this filter would be a lowpass type by inspection of the coefficients, since a constant (DC) value at the input will produce that same value at the output. Also, since all coefficients are positive, it will tend to average adjacent values of the signal.

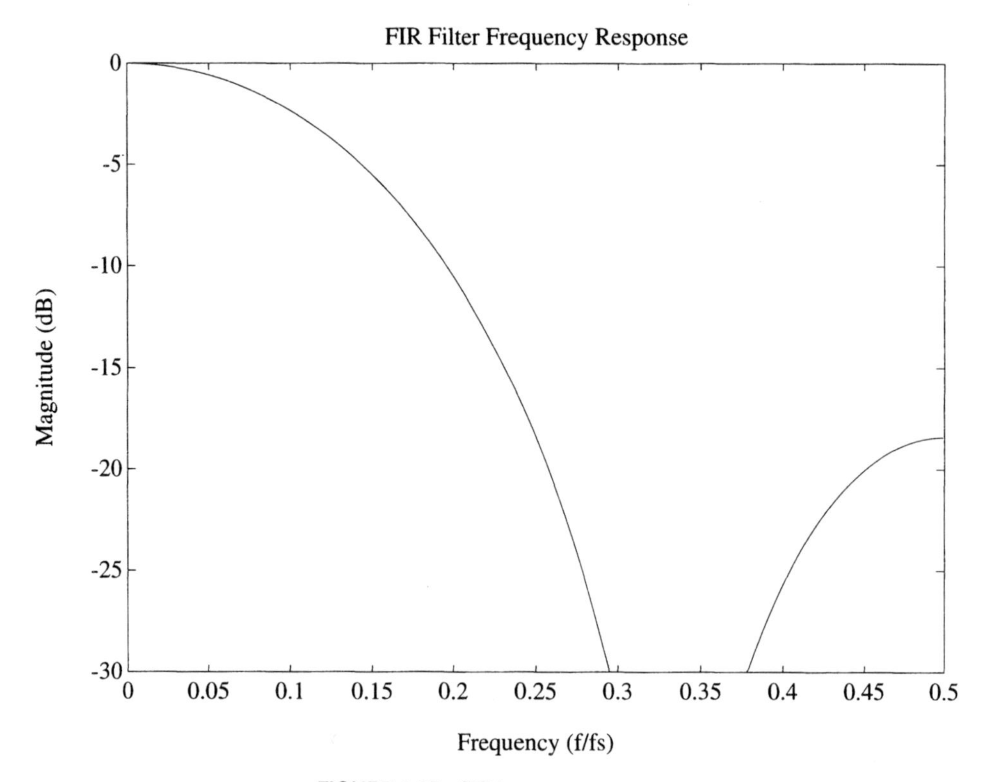

FIGURE 1.13 FIR low pass response.

The response of this FIR filter is shown in Figure 1.13. It is indeed lowpass and the nulls in the stop band are characteristic of discrete time filters in general.

As an example of the simplest IIR filter, take the following difference equation:

$$y(n) = x(n) + y(n-1).$$

Some contemplation of this filter's response to some simple inputs (like constant values, 0, 1, and so on) will lead to the conclusion that it is an integrator. For zero input, the output holds at a constant value forever. For any constant positive input greater than zero, the output grows linearly with time. For any constant negative input, the output decreases linearly with time. The frequency response of this filter is shown in Figure 1.14.

1.3.4 Filter Specifications

As mentioned previously, filters are generally specified by their performance in the frequency domain, both amplitude and phase response as a function of frequency. Figure 1.15 shows a lowpass filter magnitude response characteristic. The filter gain has

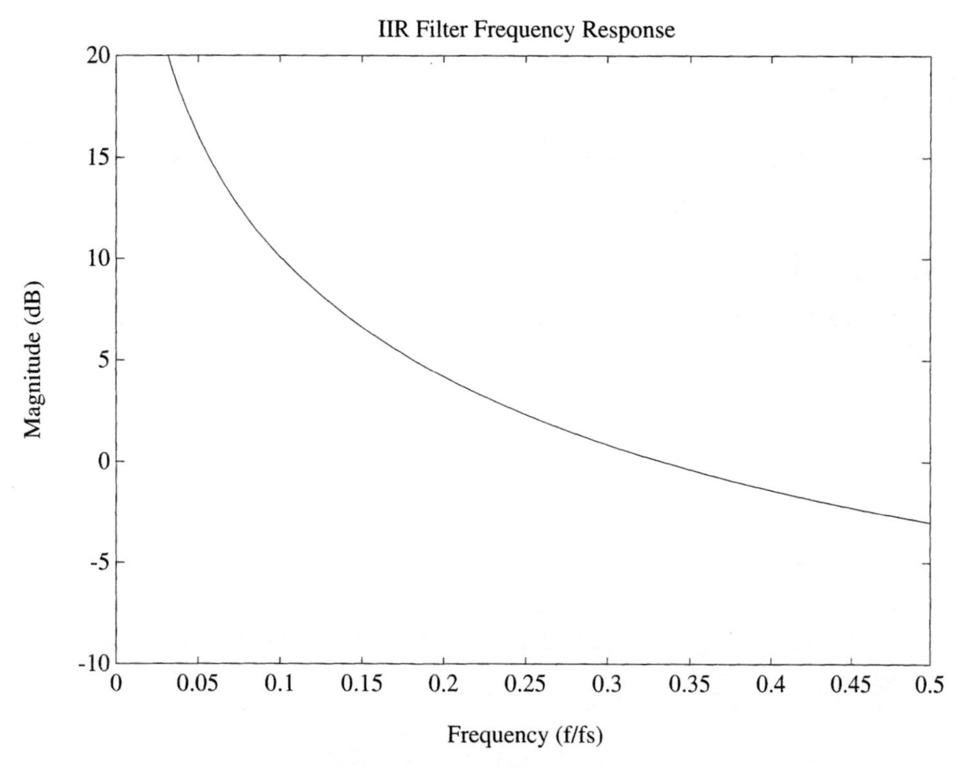

FIGURE 1.14 IIR integrator response.

been normalized to be roughly 1.0 at low frequencies and the sampling rate is normalized to unity. The figure illustrates the most important terms associated with filter specifications.

The region where the filter allows the input signal to pass to the output with little or no attenuation is called the *passband.* In a lowpass filter, the passband extends from frequency $f = 0$ to the start of the transition band, marked as frequency f_{pass} in Figure 1.15. The *transition band* is that region where the filter smoothly changes from passing the signal to stopping the signal. The end of the transition band occurs at the stopband frequency, f_{stop}. The *stopband* is the range of frequencies over which the filter is specified to attenuate the signal by a given factor. Typically, a filter will be specified by the following parameters:

(1) Passband ripple—2δ in the figure.

(2) Stopband attenuation—$1/\lambda$.

(3) Transition start and stop frequencies—f_{pass} and f_{stop}.

(4) Cutoff frequency—f_{pass}. The frequency at which the filter gain is some given factor lower than the nominal passband gain. This may be –1 dB, –3 dB or other gain value close to the passband gain.

Computer programs that calculate filter coefficients from frequency domain magnitude response parameters use the above list or some variation as the program input.

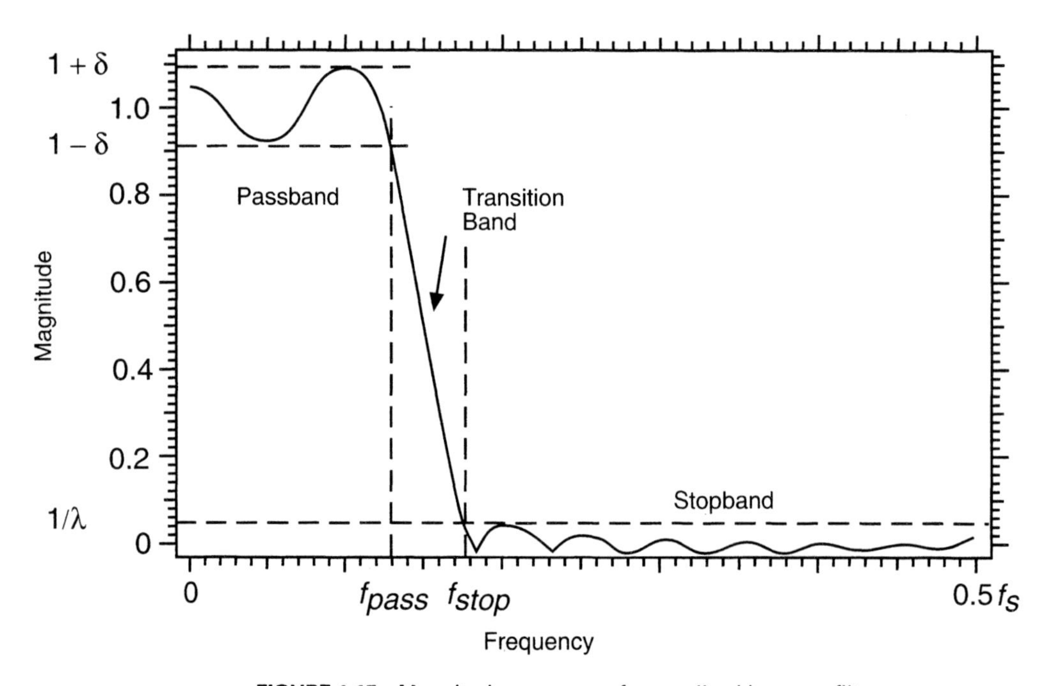

FIGURE 1.15 Magnitude response of normalized lowpass filter.

1.4 DISCRETE FOURIER TRANSFORMS

So far, the *Fourier transform* has been used several times to develop the characteristics of sequences and linear operators. The Fourier transform of a causal sequence is:

$$\mathcal{F}\{x(n)\} = X(f) = \sum_{n=0}^{\infty} x(n)e^{-j2\pi fn} \tag{1.62}$$

where the sample time period has been normalized to 1 ($T = 1$). If the sequence is of limited duration (as must be true to be of use in a computer) then

$$X(f) = \sum_{n=0}^{N-1} x(n)e^{-j2\pi fn} \tag{1.63}$$

where the sampled time domain waveform is N samples long. The inverse Fourier transform is

$$\mathcal{F}^{-1}\{X(f)\} = x(n) = \int_{-1/2}^{1/2} X(f)e^{-j2\pi fn}\,df \tag{1.64}$$

since $X(f)$ is periodic with period $1/T = 1$, the integral can be taken over any full period. Therefore,

$$x(n) = \int_{0}^{1} X(f)e^{-j2\pi fn}\,df. \tag{1.65}$$

1.4.1 Form

These representations for the Fourier transform are accurate but they have a major drawback for digital applications—the frequency variable is continuous, not discrete. To overcome this problem, both the time and frequency representations of the signal must be approximated.

To create a *discrete Fourier transform* (DFT) a sampled version of the frequency waveform is used. This sampling in the frequency domain is equivalent to convolution in the time domain with the following time waveform:

$$h_1(t) = \sum_{r=-\infty}^{\infty} \delta(t - rT).$$

This creates duplicates of the sampled time domain waveform that repeats with period T. This T is equal to the T used above in the time domain sequence. Next, by using the same number of samples in one period of the repeating frequency domain waveform as in one period of the time domain waveform, a DFT pair is obtained that is a good approximation to the continuous variable Fourier transform pair. The forward discrete Fourier transform is

$$X(k) = \sum_{n=0}^{N-1} x(n)e^{-j2\pi kn/N} \tag{1.66}$$

and the inverse discrete Fourier transform is

$$x(n) = \frac{1}{N}\sum_{k=0}^{N-1} X(k)e^{-j2\pi kn/N}. \tag{1.67}$$

For a complete development of the DFT by both graphical and theoretical means, see the text by Brigham (chapter 6).

1.4.2 Properties

This section describes some of the properties of the DFT. The corresponding paragraph numbers in the book *The Fast Fourier Transform* by Brigham (1974) are indicated. Due to the sampling theorem it is clear that no frequency higher than $1/2T$ can be represented by $X(k)$. However, the values of k extend to $N–1$, which corresponds to a frequency nearly equal to the sampling frequency $1/T$. This means that for a real sequence, the values of k from $N/2$ to $N–1$ are aliased and, in fact, the amplitudes of these values of $X(k)$ are

$$|X(k)| = |X(N-k)|, \text{ for } k = N/2 \text{ to } N-1. \tag{1.68}$$

This corresponds to Properties 8-11 and 8-14 in Brigham.

The DFT is a linear transform as is the z-transform so that the following relationships hold:

If

$$x(n) = \alpha\, a(n) + \beta\, b(n),$$

where α and β are constants, then

$$X(k) = \alpha\, A(k) + \beta\, B(k),$$

where $A(k)$ and $B(k)$ are the DFTs of the time functions $a(n)$ and $b(n)$, respectively. This corresponds to Property 8-1 in Brigham.

The DFT also displays a similar attribute under time shifting as the z-transform. If $X(k)$ is the DFT of $x(n)$ then

$$\text{DFT}\{x(n-p)\} = \sum_{n=0}^{N-1} x(n-p)e^{-j2kn/N}$$

Now define a new variable $m = r - p$ so that $n = m + p$. This gives

$$\text{DFT}\{x(n-p)\} = \sum_{m=-p}^{m=N-1-p} x(m)e^{-j\pi km/N}e^{-j2\pi kp/N},$$

which is equivalent to the following:

$$\text{DFT}\{x(n-p)\} = e^{-j2\pi kp/N}X(k). \tag{1.69}$$

This corresponds to Property 8-5 in Brigham. Remember that for the DFT it is assumed that the sequence $x(m)$ goes on forever repeating its values based on the period $n = 0$ to $N - 1$. So the meaning of the negative time arguments is simply that

$$x(-p) = x(N - p), \text{ for } p = 0 \text{ to } N - 1.$$

1.4.3 Power Spectrum

The DFT is often used as an analysis tool for determining the spectra of input sequences. Most often the amplitude of a particular frequency component in the input signal is desired. The DFT can be broken into amplitude and phase components as follows:

$$X(f) = X_{\text{real}}(f) + j\,X_{\text{imag}}(f) \tag{1.70}$$

$$X(f) = |X(f)|e^{j\theta(f)} \tag{1.71}$$

$$\text{where } |X(f)| = \sqrt{X_{\text{real}}^2 + X_{\text{imag}}^2}$$

$$\text{and } \theta(f) = \tan^{-1}\left[\frac{X_{\text{imag}}}{X_{\text{real}}}\right].$$

The power spectrum of the signal can be determined using the signal spectrum times its conjugate as follows:

$$X(k)X^*(k) = |X(k)|^2 = X_{\text{real}}^2 + X_{\text{imag}}^2. \tag{1.72}$$

There are some problems with using the DFT as a spectrum analysis tool, however. The problem of interest here concerns the assumption made in deriving the DFT that the sequence was a single period of a periodically repeating waveform. For almost all sequences there will be a discontinuity in the time waveform at the boundaries between these pseudo periods. This discontinuity will result in very high-frequency components in the resulting waveform. Since these components can be much higher than the sampling theorem limit of $1/2T$ (or half the sampling frequency) they may be aliased into the middle of the spectrum developed by the DFT.

The technique used to overcome this difficulty is called *windowing*. The problem to be overcome is the possible discontinuity at the edges of each period of the waveform. Since for a general purpose DFT algorithm there is no way to know the degree of discontinuity at the boundaries, the windowing technique simply reduces the sequence amplitude at the boundaries. It does this in a gradual and smooth manner so that no new discontinuities are produced, and the result is a substantial reduction in the aliased frequency components. This improvement does not come without a cost. Because the window is modifying the sequence before a DFT is performed, some reduction in the fidelity of the spectral representation must be expected. The result is somewhat reduced resolution of closely spaced frequency components. The best windows achieve the maximum reduction of *spurious* (or aliased) signals with the minimum degradation of spectral resolution.

There are a variety of windows, but they all work essentially the same way:

Attenuate the sequence elements near the boundaries (near $n = 0$ and $n = N - 1$) and compensate by increasing the values that are far away from the boundaries. Each window has its own individual transition from the center region to the outer elements. For a comparison of window performance see the references listed at the end of this chapter. (For example, see Harris (1983)).

1.4.4 Averaged Periodograms

Because signals are always associated with noise—either due to some physical attribute of the signal generator or external noise picked up by the signal source—the DFT of a single sequence from a continuous time process is often not a good indication of the true spectrum of the signal. The solution to this dilemma is to take multiple DFTs from successive sequences from the same signal source and take the time average of the power spectrum. If a new DFT is taken each NT seconds and successive DFTs are labeled with superscripts:

$$\text{Power Spectrum} = \sum_{i=0}^{M-1} \left[X^i_{\text{real}})^2 + (X^i_{\text{imag}})^2 \right]. \tag{1.73}$$

Clearly, the spectrum of the signal cannot be allowed to change significantly during the interval $t = 0$ to $t = \text{M}\,(NT)$.

1.4.5 The Fast Fourier Transform (FFT)

The *fast Fourier transform* (or FFT) is a very efficient algorithm for computing the DFT of a sequence. It takes advantage of the fact that many computations are repeated in the DFT due to the periodic nature of the discrete Fourier kernel: $e^{-j2\pi kn / N}$. The form of the DFT is

$$X(k) = \sum_{n=0}^{N-1} x(n) e^{-j2\pi kn/N}. \tag{1.74}$$

By letting $W^{nk} = e^{-j2\pi kn / N}$ Equation (1.74) becomes

$$X(k) = \sum_{n=0}^{N-1} x(n) W^{nk}. \tag{1.75}$$

Now, $W^{(N + qN)(k + rN)} = W^{nk}$ for all q, r that are integers due to the periodicity of the Fourier kernel.

Next break the DFT into two parts as follows:

$$X(k) = \sum_{n=0}^{N/2-1} x(2n) W_N^{2nk} + \sum_{n=0}^{N/2-1} x(2n+1) W_N^{(2n+1)k}, \tag{1.76}$$

where the subscript N on the Fourier kernel represents the size of the sequence.

By representing the even elements of the sequence $x(n)$ by x_{ev} and the odd elements by x_{od}, the equation can be rewritten

$$X(k) = \sum_{n=0}^{N/2-1} x_{ev}(n)W_{N/2}^{nk} + W_{N/2}^{k} \sum_{n=0}^{N/2-1} x_{od}(n)W_{N/2}^{nk}. \tag{1.77}$$

Now there are two expressions in the form of DFTs so Equation (1.77) can be simplified as follows:

$$X(k) = X_{ev}(n) + W_{N/2}^{k} X_{od}(n). \tag{1.78}$$

Notice that only DFTs of $N/2$ points need be calculated to find the value of $X(k)$. Since the index k must go to $N-1$, however, the periodic property of the even and odd DFTs is used. In other words,

$$X_{ev}(k) = X_{ev}(k - \tfrac{N}{2}) \qquad \text{for } \tfrac{N}{2} \le k \le N-1. \tag{1.79}$$

The process of dividing the resulting DFTs into even and odd halves can be repeated until one is left with only two point DFTs to evaluate

$$\begin{aligned} \Lambda(k) &= \lambda(0) + \lambda(1)e^{-j2\pi k/2} && \text{for all } k \\ &= \lambda(0) + \lambda(1) && \text{for } k \text{ even} \\ &= \lambda(0) - \lambda(1) && \text{for } k \text{ odd.} \end{aligned}$$

Therefore, for 2 point DFTs no multiplication is required, only additions and subtractions. To compute the complete DFT still requires multiplication of the individual 2-point DFTs by appropriate factors of W ranging from W^0 to $W^{N/2-1}$. Figure 1.16 shows a flow graph of a complete 32-point FFT. The savings in computation due to the FFT algorithm is as follows.

For the original DFT, N complex multiplications are required for each of N values of k. Also, $N-1$ additions are required for each k.

In an FFT each function of the form

$$\lambda(0) \pm W^{P}\lambda(1)$$

(called a *butterfly* due to its flow graph shape) requires one multiplication and two additions. From the flow graph in Figure 1.16 the number of butterflies is

$$\text{Number of butterflies} = \frac{N}{2}\log_2(N).$$

This is because there are $N/2$ rows of butterflies (since each butterfly has two inputs) and there are $\log_2(N)$ columns of butterflies.

Table 1.1 gives a listing of additions and multiplications for various sizes of FFTs and DFTs. The dramatic savings in time for larger DFTs provided in the FFT has made this method of spectral analysis practical in many cases where a straight DFT computa-

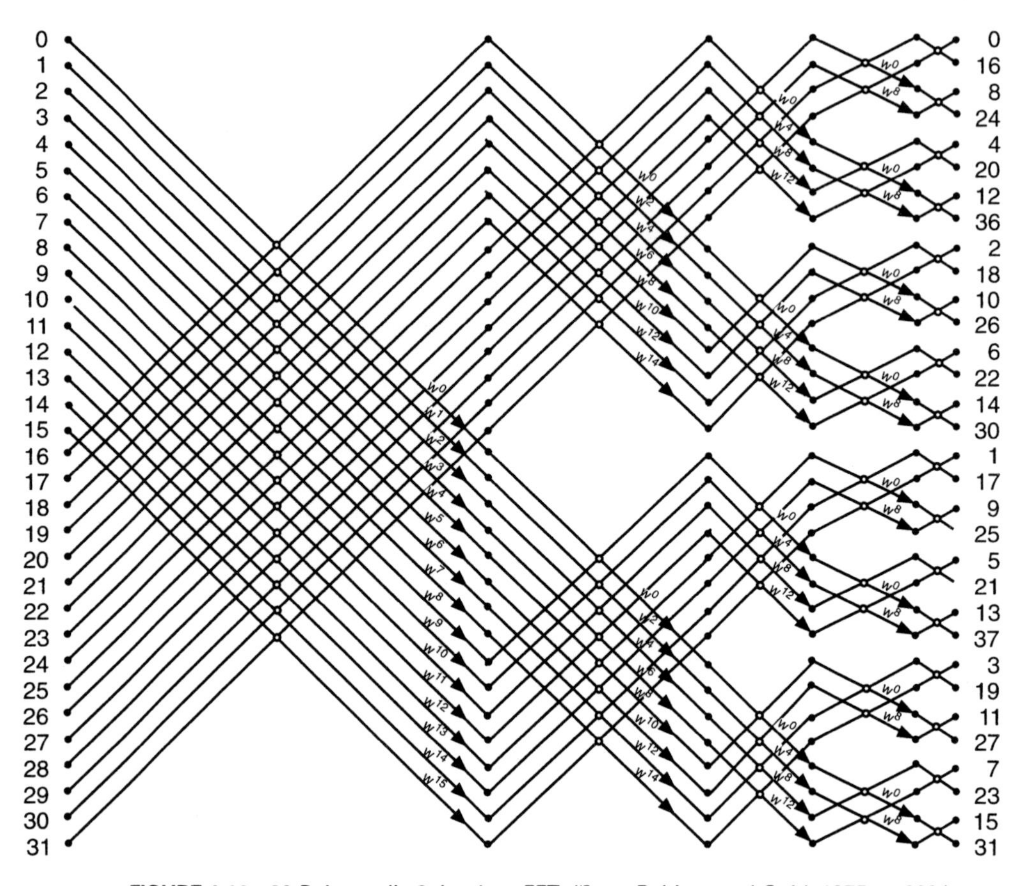

FIGURE 1.16 32-Point, radix 2, in-place FFT. (From Rabiner and Gold, 1975, p. 380.)

tion would be much too time consuming. Also, the FFT can be used for performing operations in the frequency domain that would require much more time consuming computations in the time domain.

1.4.6 An Example of the FFT

In order to help the reader gain more understanding of spectrum analysis with the FFT, a simple example is presented here. An input signal to a 16-point FFT processor is as follows:

$$x(n) = \cos[2\pi\,(4n/16)].$$

The argument of the cosine has been written in an unusual way to emphasize the frequency of the waveform when processed by a 16-point FFT. The amplitude of this signal is 1.0 and it is clearly a real signal, the imaginary component having zero amplitude. Figure 1.17 shows the 16 samples that comprise $x(0)$ to $x(15)$.

TABLE 1.1 Comparison of Number of Butterfly Operations in the DFT and FFT, (each operation is one complex multiply/accumulate calculation).

Transform Length (N)	DFT Operations (N^2)	FFT Operations $N\text{LOG}_2(N)$
8	64	24
16	256	64
32	1024	160
64	4096	384
128	16384	896
256	65536	1024
512	262144	4608
1024	1048576	10240
2048	4194304	22528

With this input a 16-point FFT will produce a very simple output. This output is shown in Figure 1.18. It is a spike at $k = 4$ of amplitude 0.5 and a spike at $k = 12$ of amplitude -0.5. The spike nature in the FFT output in this example occurs because for a cosine waveform of arbitrary frequency the Fourier transform is

$$X(f) = \int_{-\infty}^{+\infty} \cos(2\pi f_0 t) e^{-j2\pi ft}\, dt.$$

Representing the cosine by exponentials

$$X(f) = \frac{1}{2}\int_{-\infty}^{+\infty} e^{j2\pi(f_0 - f)t}\, dt - \frac{1}{2}\int_{-\infty}^{+\infty} e^{-j2\pi(f_0 + f)t}\, dt.$$

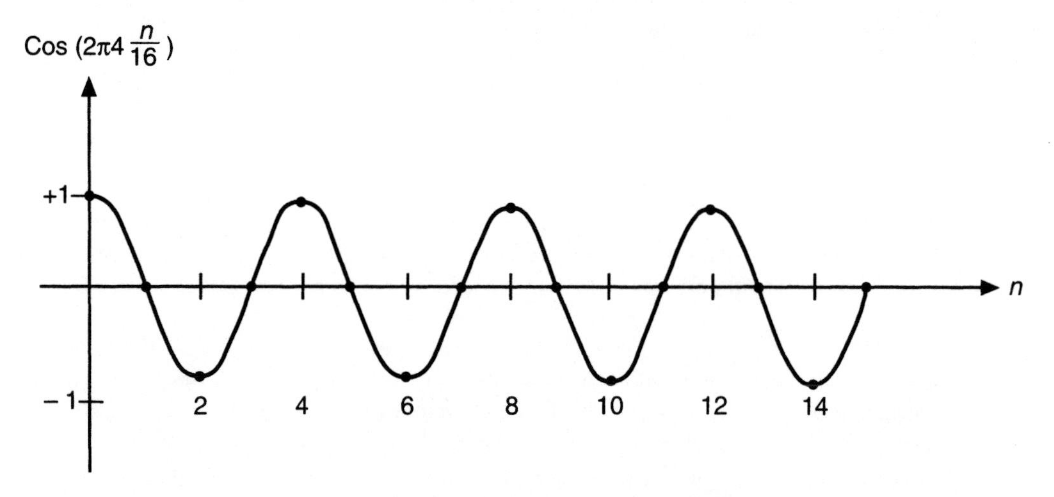

FIGURE 1.17 Input to 16 point FFT.

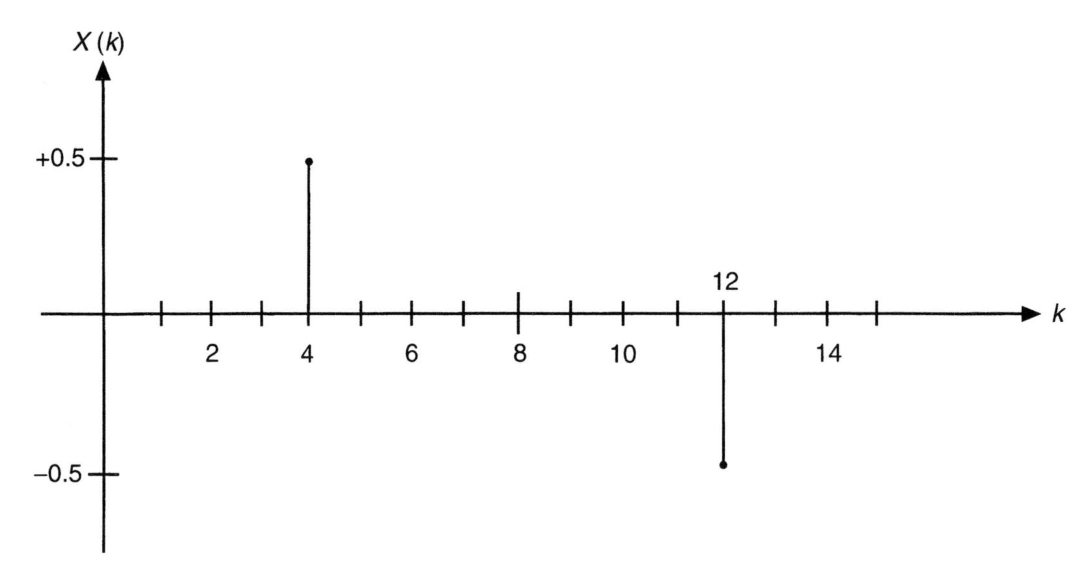

FIGURE 1.18 Output of 16-point FFT.

It can be shown that the integrand in the two integrals above integrates to 0 unless the argument of the exponential is 0. If the argument of the exponential is zero, the result is two infinite spikes, one at $f = f_0$ and the other at $f = -f_0$. These are delta functions in the frequency domain.

Based on these results, and remembering that the impulse sequence is the digital analog of the delta function, the results for the FFT seem more plausible. It is still left to explain why $k = 12$ should be equivalent to $f = -f_0$. Referring back to the development of the DFT, it was necessary at one point for the frequency spectrum to become periodic with period f_s. Also, in the DFT only positive indices are used. Combining these two facts one can obtain the results shown in Figure 1.18.

1.5 NONLINEAR OPERATORS

Most of this book is devoted to linear operators and linear-signal processing because these are the most commonly used techniques in DSP. However, there are several nonlinear operators that are very useful in one-dimensional DSP. This section introduces the simple class of nonlinear operators that compress or clip the input to derive the output sequence.

There is often a need to reduce the number of significant bits in a quantized sequence. This is sometimes done by truncation of the least significant bits. This process is advantageous because it is linear: The quantization error is increased uniformly over the entire range of values of the sequence. There are many applications, however, where the need for accuracy in quantization is considerably less at high-signal values than at low-signal values. This is true in telephone voice communications where the human ear's ability to differentiate between amplitudes of sound waves decreases with the amplitude of the sound. In these cases, a nonlinear function is applied to the signal and the resulting output range of values is quantized uniformly with the available bits.

This process is illustrated in Figure 1.19. First, the input signal is shown in Figure 1.19(a). The accuracy is 12 bits and the range is 0 to 4.095 volts, so each quantization level represents 1 mV. It is necessary because of some system consideration (such as transmission bandwidth) to reduce the number bits in each word to 8. Figure 1.19(b) shows that the resulting quantization levels are 16 times as coarse. Figure 1.19(c) shows the result of applying a linear-logarithmic compression to the input signal. In this type of compression the low-level signals (out to some specified value) are unchanged from the input values. Beginning at a selected level, say $f_{in} = a$, a logarithmic function is applied. The form of the function might be

$$f_{\text{out}} = a + A\log_{10}(1 + f_{\text{in}} - a)$$

so that at $f_{in} = a$ the output also equals a and A is chosen to place the maximum value of f_{out} at the desired point.

A simpler version of the same process is shown in Figure 1.20. Instead of applying a logarithmic function from the point $f = a$ onward, the output values for $f \geq a$ are all the same. This is an example of clipping. A region of interest is defined and any values outside the region are given a constant output.

1.5.1 μ-Law and A-Law Compression

There are two other compression laws worth listing because of their use in telephony—the *μ-law* and *A-law* conversions. The *μ-law* conversion is defined as follows:

$$f_{\text{out}} = \text{sgn}(f_{\text{in}})\frac{\ln(1 + \mu|f_{\text{in}}|)}{\ln(1 + \mu)}, \tag{1.80}$$

where sgn() is a function that takes the sign of its argument, and μ is the compression parameter (255 for North American telephone transmission). The input value f_{in} must be normalized to lie between −1 and +1. The A-law conversion equations are as follows:

$$f_{\text{out}} = \text{sgn}(f_{\text{in}})\frac{A|f_{\text{in}}|}{1 + \ln(A)}$$

for $|f_{\text{in}}|$ between 0 and $1/A$ and

$$f_{\text{out}} = \text{sgn}(f_{\text{in}})\frac{1 + \ln(A|f_{\text{in}}|)}{1 + \ln(A)} \tag{1.81}$$

for $|f_{\text{in}}|$ between $1/A$ and 1.

In these equations, A is the compression parameter (87.6 for European telephone transmission).

An extreme version of clipping is used in some applications of image processing to produce binary pictures. In this technique a threshold is chosen (usually based on a his-

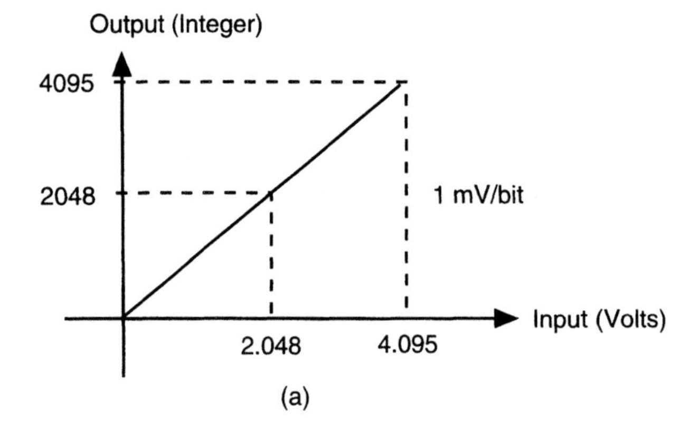

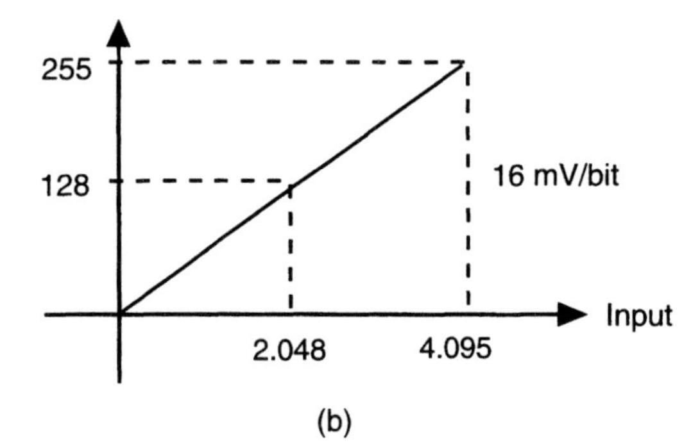

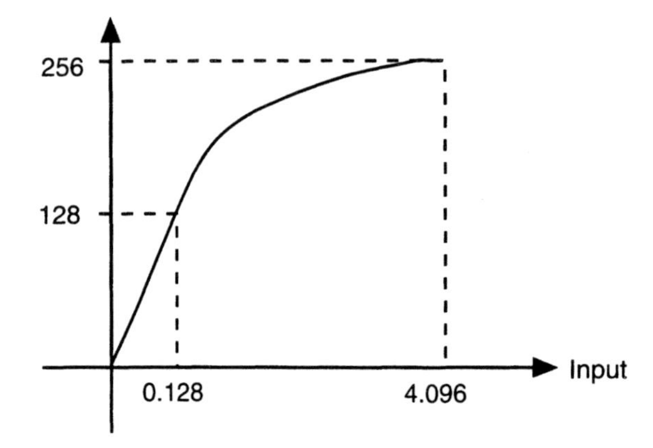

FIGURE 1.19 **(a)** Linear 12-bit ADC. **(b)** Linear 8-bit ADC. **(c)** Nonlinear conversion.

togram of the picture elements) and any image element with a value higher than threshold is set to 1 and any element with a value lower than threshold is set to zero. In this way the significant bits are reduced to only one. Pictures properly thresholded can produce excellent outlines of the most interesting objects in the image, which simplifies further processing considerably.

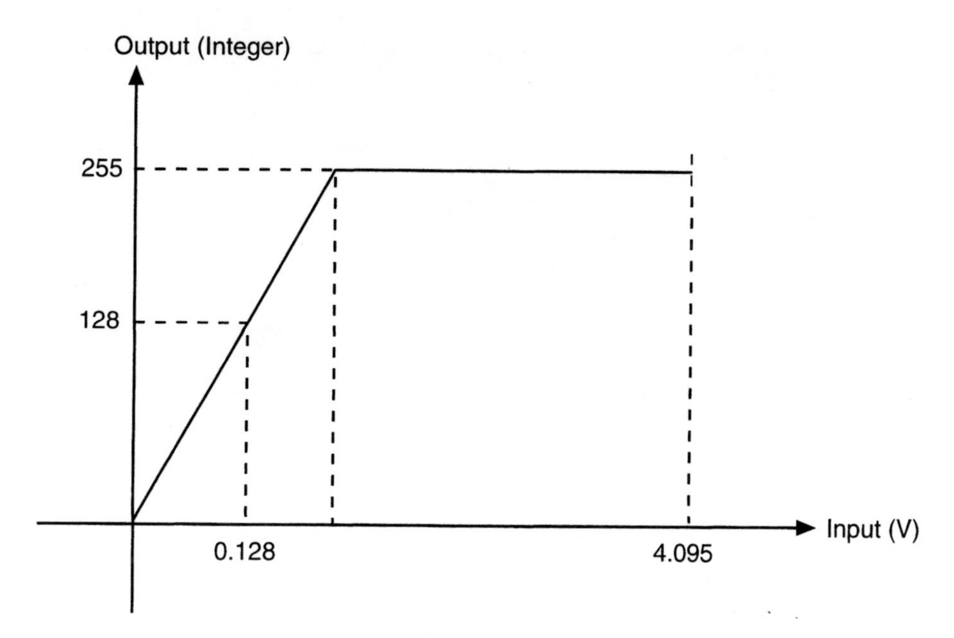

FIGURE 1.20 Clipping to 8 bits.

1.6 PROBABILITY AND RANDOM PROCESSES

The signals of interest in most signal-processing problems are embedded in an environment of noise and interference. The noise may be due to spurious signals picked up during transmission (interference), or due to the noise characteristics of the electronics that receives the signal or a number of other sources. To deal effectively with noise in a signal, some model of the noise or of the signal plus noise must be used. Most often a probabilistic model is used, since the noise is, by nature, unpredictable. This section introduces the concepts of probability and randomness that are basic to digital signal processing and gives some examples of the way a composite signal of interest plus noise is modeled.

1.6.1 Basic Probability

Probability begins by defining the probability of an event labeled A as $P(A)$. Event A can be the result of a coin toss, the outcome of a horse race, or any other result of an activity that is not completely predictable. There are three attributes of this probability $P(A)$:

(1) $P(A) >= 0$. This simply means that any result will either have a positive chance of occurrence or no chance of occurrence.

(2) P (All possible outcomes) = 1. This indicates that some result among those possible is bound to occur, a probability of 1 being certainty.

(3) For $\{A_i\}$, where $(A_i \cap A_i) = 0$, $P(\cup A_i) = \Sigma_i P(A_i)$. For a set of events, $\{A_i\}$, where the events are mutually disjoint (no two can occur as the result of a single trial of the activity), the probability of any one of the events occurring is equal to the sum of their individual probabilities.

With probability defined in this way, the discussion can be extended to joint and conditional probabilities. Joint probability is defined as the probability of occurrence of a specific set of two or more events as the result of a single trial of an activity. For instance, the probability that horse A will finish third and horse B will finish first in a particular horse race is a joint probability. This is written:

$$P(A \cap B) = P(A \text{ and } B) = P(AB). \tag{1.82}$$

Conditional probability is defined as the probability of occurrence of an event A given that B has occurred. The probability assigned to event A is conditioned by some knowledge of event B. This is written

$$P(A \text{ given } B) = P(A|B). \tag{1.83}$$

If this conditional probability, $P(A|B)$, and the probability of B are both known, the probability of both of these events occurring (joint probability) is

$$P(AB) = P(A|B)P(B). \tag{1.84}$$

So the conditional probability is multiplied by the probability of the condition (event B) to get the joint probability. Another way to write this equation is

$$P(A|B) = \frac{P(AB)}{P(B)} \tag{1.85}$$

This is another way to define conditional probability once joint probability is understood.

1.6.2 Random Variables

In signal processing, the probability of a signal taking on a certain value or lying in a certain range of values is often desired. The signal in this case can be thought of as a *random variable* (an element whose set of possible values is the set of outcomes of the activity). For instance, for the random variable X, the following set of events, which could occur, may exist:

Event A X takes on the value of 5 ($X = 5$)

Event B $X = 19$

Event C $X = 1.66$

etc.

This is a useful set of events for discrete variables that can only take on certain specified values. A more practical set of events for continuous variables associates each event with the variable lying within a range of values. A cumulative distribution function (or CDF) for a random variable can be defined as follows:

$$F(x) = P(X \leq x). \tag{1.86}$$

This cumulative distribution, function, then, is a monotonically increasing function of the independent variable x and is valid only for the particular random variable, X. Figure 1.21 shows an example of a distribution function for a random variable. If $F(x)$ is differentiated with respect to x the probability density function (or PDF) for X is obtained, represented as follows:

$$p(x) = \frac{dF(x)}{dx}. \tag{1.87}$$

Integrating $p(x)$ gives the distribution function back again as follows:

$$F(x) = \int_{-\infty}^{x} p(\lambda)d\lambda. \tag{1.88}$$

Since $F(x)$ is always monotonically increasing, $p(x)$ must be always positive or zero. Figure 1.22 shows the density function for the distribution of Figure 1.21. The utility of these functions can be illustrated by determining the probability that the random variable X lies between a and b. By using probability Property 3 from above

$$P(X \leq b) = P(a < X \leq b) + P(X \leq a). \tag{1.89}$$

This is true because the two conditions on the right-hand side are independent (mutually exclusive) and X must meet one or the other if it meets the condition on the left-hand side. This equation can be expressed using the definition of the distribution:

$$\begin{aligned} P(a < X \leq b) &= F(b) - F(a) \\ &= \int_{a}^{b} p(x)dx. \end{aligned} \tag{1.90}$$

In this way, knowing the distribution or the density function allows the calculation of the probability that X lies within any given range.

1.6.3 Mean, Variance, and Gaussian Random Variables

There is an operator in random variable theory called the *expectation operator* usually written $E[x]$. This expression is pronounced "the expected value of x." The expectation operator extracts from a random variable the value that the variable is most likely to take

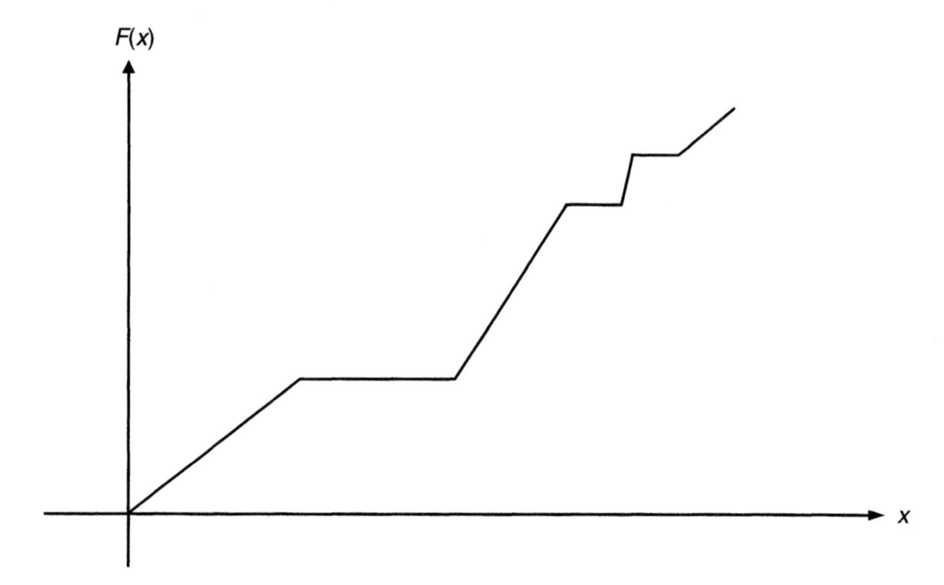

FIGURE 1.21 An example of cumulative distribution function (CDF).

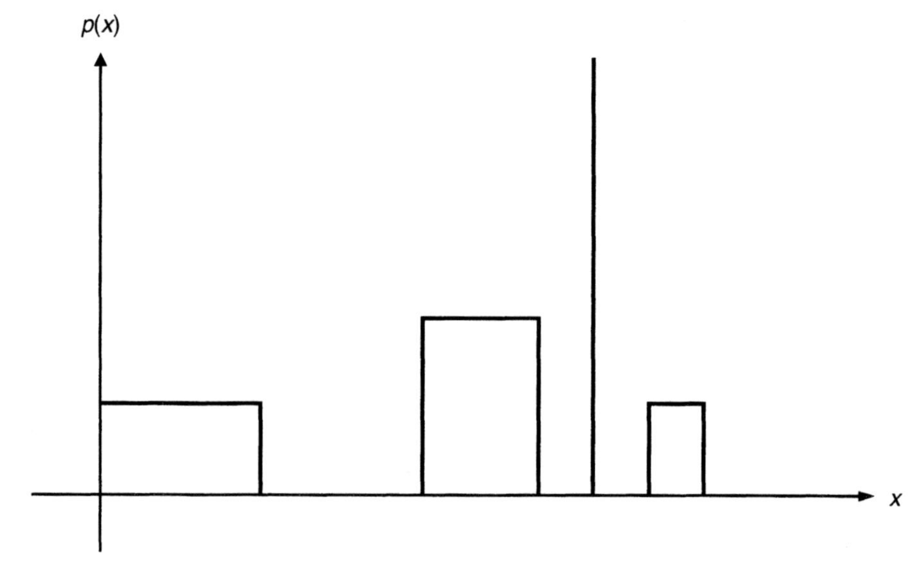

FIGURE 1.22 Density function.

on. The expected value is sometimes called the *mean, average,* or *first moment* of the variable and is calculated from the density function as follows:

$$E[x] = \int_{-\infty}^{\infty} xp(x)dx. \tag{1.91}$$

A typical density function for a random variable is shown in Figure 1.23. The most likely value of variable x is also indicated in the figure. The expected value can be thought of as a "center of gravity" or first moment of the random variable x.

The variance of a random variable is defined as

$$\sigma^2 = \text{Var}\{x\} = E\left[(x - E[x])^2\right], \tag{1.92}$$

where σ is the root mean square value of the variable's difference from the mean. The variance is sometimes called the *mean square value* of x.

By extending the use of the expectation operator to joint probability densities, a variable Y can be a function of two random variables, s and t such that

$$Y = \mathcal{O}\{s, t\}.$$

Then the expected value of Y will be

$$E[Y] = \int_{-\infty}^{\infty}\int_{-\infty}^{\infty} \mathcal{O}[s, t\}p(s, t)dsdt \tag{1.93}$$

where the joint probability density of s and t ($p(s,t)$), is required in the equation. The correlation of two random variables is defined to be the expected value of their product

$$E[st\} = \int_{-\infty}^{\infty}\int_{-\infty}^{\infty} st\ p(s, t)dsdt. \tag{1.94}$$

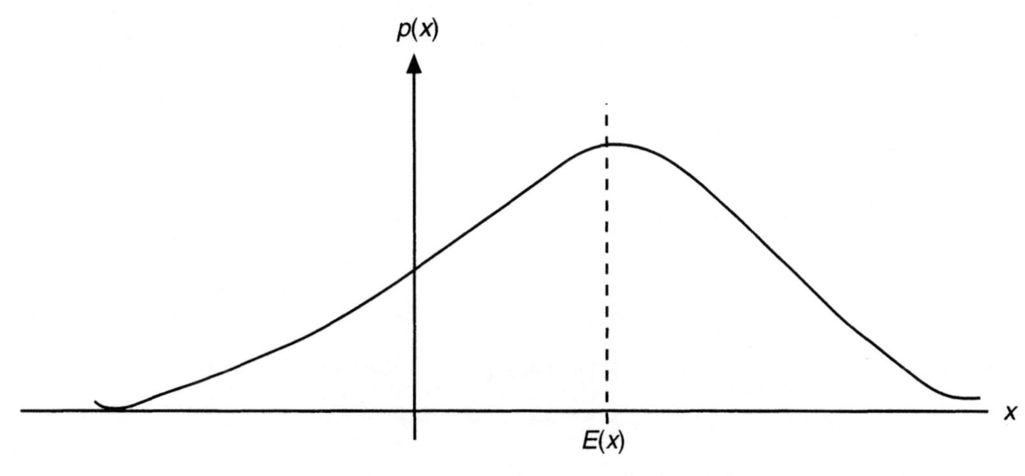

FIGURE 1.23 Continuous PDF showing $E[x]$.

This definition will be used in the development of autocorrelation in section 1.6.5.

There is a set of random variables called *Gaussian random variables* whose density functions have special characteristics that make them particularly easy to analyze. Also, many physical processes give rise to approximately this sort of density function. A *Gaussian density function* has the following form:

$$p(x) = \frac{1}{\sqrt{2\pi}\sigma} \exp\left[-\frac{(x-\mu)^2}{2\sigma^2} \right], \tag{1.95}$$

where μ is the mean value of x and σ^2 is the variance.

1.6.4 Quantization of Sequences

Quantization is to the amplitude domain of a continuous analog signal as sampling is to the time domain. It is the step that allows a continuous amplitude signal to be represented in the discrete amplitude increments available in a digital computer. To analyze the process of quantization, it is useful to diagram a system as shown in Figure 1.24. The illustration shows a continuous amplitude input signal, f, which is sampled and quantized, then reconstructed in the continuous amplitude domain. The output signal is $\hat{f}$. By comparing the input and output of this process the effect of quantization can be illustrated.

The action of the box marked quantization in Figure 1.24 is illustrated in Figure 1.25. A set of decision levels is applied to each input signal, and the two levels which bracket the signal above and below are determined. A digital code is assigned to the region between each levels. In Figure 1.25, the digital code consists of 6 bits and runs from binary 0 to binary 63. The application of these decision levels and the assignment of a code to the input signal sample is the complete process of quantization. Reconstruction of the signal is accomplished by assigning a reconstruction level to each digital code.

The task that remains is to assign actual values to the decision levels and the reconstruction levels. Referring to Figure 1.25, the minimum value of the input signal is labeled a_L and the maximum value is labeled a_U. If the signal f has a probability density of $p(f)$, then the mean squared error due to the quantization and reconstruction process is

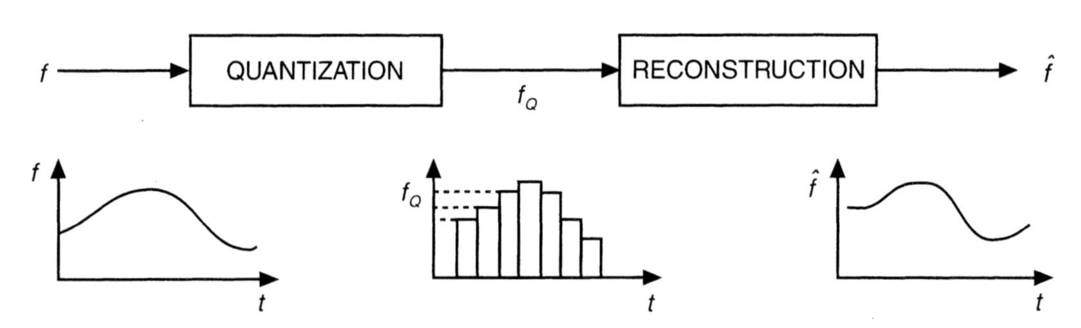

FIGURE 1.24 Quantization and reconstruction of a signal.

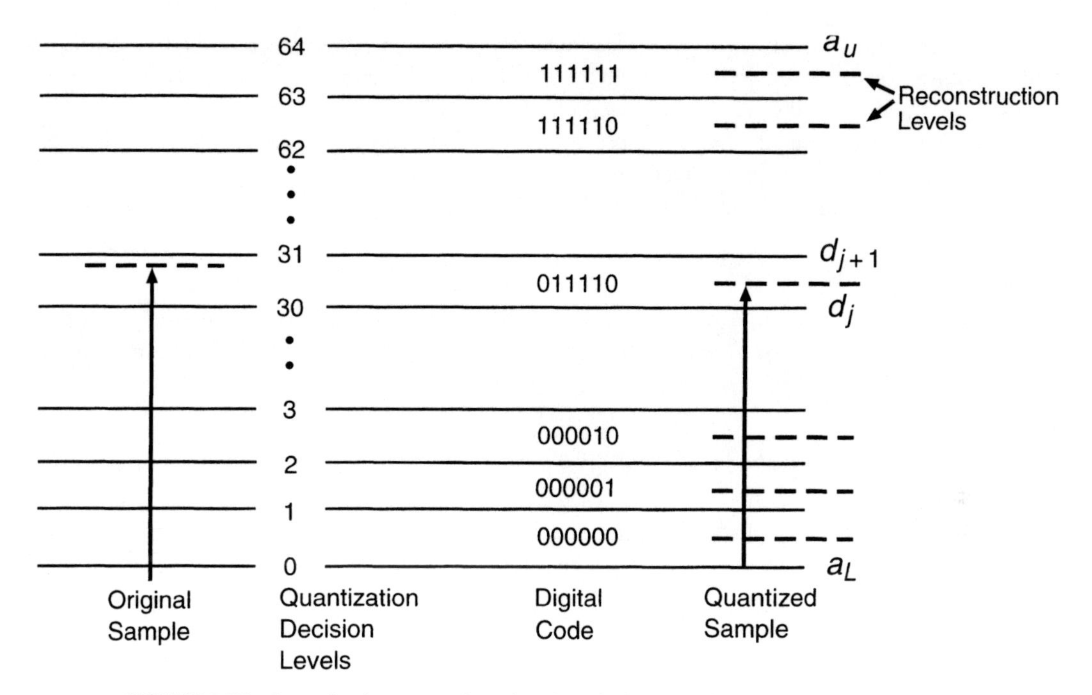

FIGURE 1.25 Quantization operation showing decision and reconstruction levels.

$$\epsilon = E\left\{(f-\hat{f})^2\right\} = \int_{a_L}^{a_U} (f-\hat{f})^2 p(f)df,$$

and if the signal range is broken up into the segments between decision levels d_j and d_{j+1}, then

$$\epsilon = E\left\{(f-\hat{f})^2\right\} = \sum_{j=0}^{J-1} \int_{d_j}^{d_{j+1}} (f-r_j)^2 p(f)df.$$

Numerical solutions can be determined that minimize ε for several common probability densities. The most common assumption is a uniform density ($p(f)$ equals $1/N$ for all values of f, where N is the number of decision intervals). In this case, the decision levels are uniformly spaced throughout the interval and the reconstruction levels are centered between decision levels. This method of quantization is almost universal in commercial analog-to-digital converters. For this case the error in the analog-to-digital converter output is uniformly distributed from $-\frac{1}{2}$ of the least significant bit to $+\frac{1}{2}$ of the least significant bit. If it is assumed that the value of the least significant bit is unity, then the mean squared error due to this uniform quantization is given by:

$$\text{var}\{\epsilon\} = \int_{-\frac{1}{2}}^{+\frac{1}{2}} (f-\hat{f})^2 p(f)df = \int_{-\frac{1}{2}}^{+\frac{1}{2}} f^2 df = \frac{1}{12},$$

since $p(f) = 1$ from $-\frac{1}{2}$ to $+\frac{1}{2}$. This mean squared error gives the equivalent variance, or noise power, added to the original continuous analog samples as a result of the uniform quantization. If it is further assumed that the quantization error can be modeled as a stationary, uncorrelated white noise process (which is a good approximation when the number of quantization levels is greater than 16), then a maximum signal-to-noise ratio (SNR) can be defined for a quantization process of B bits (2^B quantization levels) as follows:

$$\text{SNR} = 10\log_{10}(V^2 / \text{var}\{\epsilon\}) = 10\log_{10}(12V^2),$$

where V^2 is the total signal power. For example, if a sinusoid is sampled with a peak amplitude of 2^{B-1}, then $V^2 = 2^{2B}/8$ giving the signal to noise ratio for a full scale sinusoid as

$$\text{SNR} = 10\log_{10}((1.5)(2^{2B})) = 6.02B + 1.76.$$

This value of SNR is often referred to as the theoretical signal-to-noise ratio for a B bit analog-to-digital converter. Because the analog circuits in a practical analog-to-digital converter always add some additional noise, the SNR of a real-world converter is always less than this value.

1.6.5 Random Processes, Autocorrelation, and Spectral Density

A *random process* is a function composed of random variables. An example is the random process $f(t)$. For each value of t, the process $f(t)$ can be considered a random variable. For $t = a$ there is a random variable $f(a)$ that has a probability density, an expected value (or mean), and a variance as defined in section 1.6.3. In a two-dimensional image, the function would be $f(x,y)$, where x and y are spatial variables. A two-dimensional random process is usually called a *random field.* Each $f(a,b)$ is a random variable.

One of the important aspects of a random process is the way in which the random variables at different points in the process are related to each other. The concept of joint probability is extended to distribution and density functions. A *joint probability distribution* is defined as

$$F(s,t) = P(S \leq s, T \leq t) \text{ (where } S \text{ and } T \text{ are some constants)},$$

and the corresponding density function is defined as

$$p(s,t) = \frac{\partial^2 F(s,t)}{\partial s \partial t}. \tag{1.96}$$

The integral relationship between distribution and density in this case is

$$F(s,t) = \int_{-\infty}^{s}\int_{-\infty}^{t} p(\alpha,\beta)\,d\alpha\, d\beta. \tag{1.97}$$

In section 1.6.3 it was shown that the correlation of two random variables is the expected value of their product. The *autocorrelation* of a random process is the expected value of the products of the random variables which make up the process. The symbol for autocorrelation is $R_{ff}(t_1, t_2)$ for the function $f(t)$ and the definition is

$$R_{ff}(t_1, t_2) = E[f(t_1)f(t_2)] \tag{1.98}$$

$$= \int_{-\infty}^{\infty}\int_{-\infty}^{\infty} \alpha\, \beta\, p_f(\alpha, \beta; t_1, t_2)\, d\alpha\, d\beta, \tag{1.99}$$

where $p_f(\alpha, \beta; t_1, t_2)$ is the joint probability density $f(t_1)$ and $f(t_2)$. By including α and β in the parentheses the dependence of p_f on these variables is made explicit.

In the general case, the autocorrelation can have different values for each value of t_1 and t_2. However, there is an important special class of random processes called *stationary processes* for which the form of the autocorrelation is somewhat simpler. In stationary random processes, the autocorrelation is only a function of the difference between the two time variables. For stationary processes

$$R_{ff}(t_2 - t_1) = R_{ff}(\xi) = E[f(t - \xi)f(t)]. \tag{1.100}$$

In section 1.6.6 the continuous variable theory presented here is extended to discrete variables and the concept of modeling real world signals is introduced.

1.6.6 Modeling Real-World Signals with AR Processes

By its nature, a noise process cannot be specified as a function of time in the way a deterministic signal can. Usually a noise process can be described with a probability function and the first and second moments of the process. Although this is only a partial characterization, a considerable amount of analysis can be performed using moment parameters alone. The first moment of a process is simply its average or mean value. In this section, all processes will have zero mean, simplifying the algebra and derivations but providing results for the most common set of processes.

The second moment is the autocorrelation of the process

$$r(n, n-k) = E[u(n)u^*(n-k)], \qquad \text{for } k = 0, \pm 1, \pm 2, \ldots.$$

The processes considered here are *stationary* to *second order.* This means that the first and second order statistics do not change with time. This allows the autocorrelation to be represented by

$$r(n, n-k) = r(k), \qquad \text{for } k = 0, \pm 1, \pm 2, \ldots$$

since it is a function only of the time *difference* between samples and not the time variable itself. In any process, an important member of the set of autocorrelation values is $r(0)$, which is

$$r(0) = E\{u(n)u^*(n)\} = E\{|u(n)|^2\}, \tag{1.101}$$

which is the mean square value of the process. For a zero mean process this is equal to the variance of the signal

$$r(0) = \text{var}\{u\}. \tag{1.102}$$

The process can be represented by a vector $\mathbf{u}(n)$ where

$$\mathbf{u}(n) = \begin{bmatrix} u(n) \\ u(n-1) \\ u(n-2) \\ \vdots \\ u(n-M+1) \end{bmatrix} \tag{1.103}$$

Then the autocorrelation can be represented in matrix form

$$\mathbf{R} = E\left\{\mathbf{u}(n)\mathbf{u}^H(n)\right\} \tag{1.104}$$

$$\mathbf{R} = \begin{bmatrix} r(0) & r(1) & r(2)... & r(m-1) \\ r(-1) & r(0) & r(1)... & \vdots \\ r(-2) & r(-1) & r(0)... & \vdots \\ \vdots & \vdots & \vdots & r(-1) \\ \vdots & \vdots & \vdots & r(0) \\ r(-M+1) & r(-M+2) & ...... & r(1) \end{bmatrix}$$

The second moment of a noise process is important because it is directly related to the power spectrum of the process. The relationship is

$$S(f) = \sum_{k=-M+1}^{M-1} r(k)e^{-j2\pi fk}, \tag{1.105}$$

which is the discrete Fourier transform (DFT) of the autocorrelation of the process ($r(k)$). Thus, the autocorrelation is the time domain description of the second order statistics, and the power spectral density, $S(f)$, is the frequency domain representation. This power spectral density can be modified by discrete time filters.

Discrete time filters may be classified as *autoregressive* (AR), *moving average* (MA), or a *combination of the two* (ARMA). Examples of these filter structures and the z-transforms of each of their impulse responses are shown in Figure 1.26. It is theoretically possible to create any arbitrary output stochastic process from an input white noise Gaussian process using a filter of sufficiently high (possibly infinite) order.

Referring again to the three filter structures in Figure 1.26, it is possible to create any arbitrary transfer function $H(z)$ with any one of the three structures. However, the orders of the realizations will be very different for one structure as compared to another. For instance, an infinite order MA filter may be required to duplicate an Mth order AR filter.

One of the most basic theorems of adaptive and optimal filter theory is the *Wold*

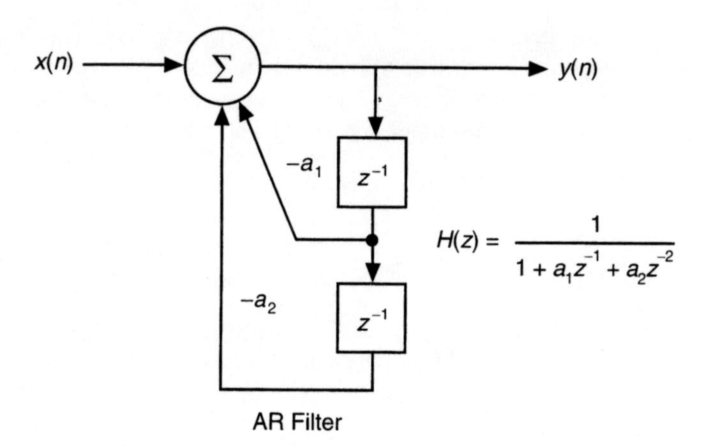

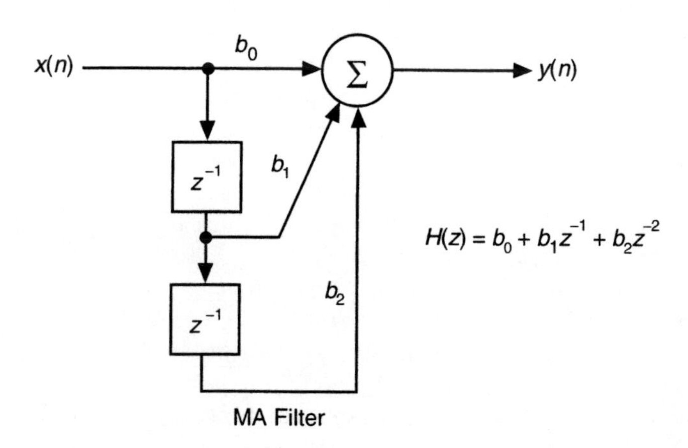

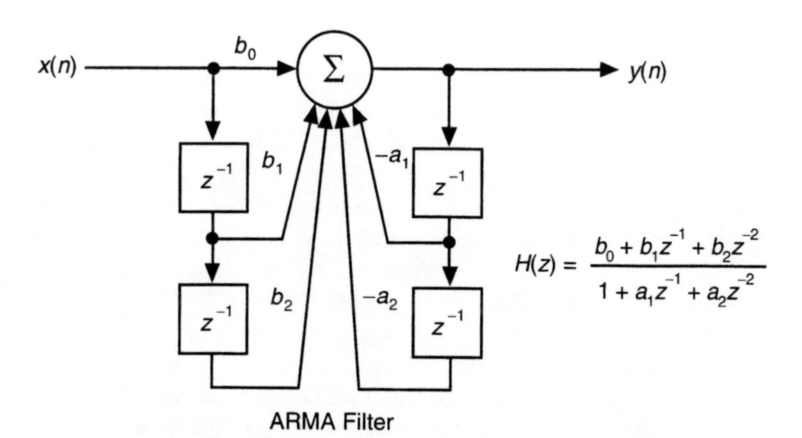

FIGURE 1.26 AR, MA, and ARMA filter structures.

decomposition. This theorem states that any real-world process can be decomposed into a deterministic component (such as a sum of sine waves at specified amplitudes, phases, and frequencies) and a noise process. In addition, the theorem states that the noise process can be modeled as the output of a linear filter excited at its input by a white noise signal.

1.7 ADAPTIVE FILTERS AND SYSTEMS

The problem of determining the optimum linear filter was solved by Norbert Wiener and others. The solution is referred to as the *Wiener filter* and is discussed in section 1.7.1. Adaptive filters and adaptive systems attempt to find an optimum set of filter parameters (often by approximating the Wiener optimum filter) based on the time varying input and output signals. In this section, adaptive filters and their application in closed loop adaptive systems are discussed briefly. Closed-loop adaptive systems are distinguished from open-loop systems by the fact that in a closed-loop system the adaptive processor is controlled based on information obtained from the input signal and the output signal of the processor. Figure 1.27 illustrates a basic adaptive system consisting of a processor that is controlled by an adaptive algorithm, which is in turn controlled by a performance calculation algorithm that has direct knowledge of the input and output signals.

Closed-loop adaptive systems have the advantage that the performance calculation algorithm can continuously monitor the input signal (d) and the output signal (y) and determine if the performance of the system is within acceptable limits. However, because several feedback loops may exist in this adaptive structure, the automatic optimization algorithm may be difficult to design, the system may become unstable or may result in nonunique and/or nonoptimum solutions. In other situations, the adaptation process may not converge and lead to a system with grossly poor performance. In spite of these possible drawbacks, closed-loop adaptive systems are widely used in communications, digital storage systems, radar, sonar, and biomedical systems.

The general adaptive system shown in Figure 1.27(a) can be applied in several ways. The most common application is prediction, where the desired signal (d) is the application provided input signal and a delayed version of the input signal is provided to the input of the adaptive processor (x) as shown in Figure 1.27(b). The adaptive processor must then try to predict the current input signal in order to reduce the error signal (ε) toward a mean squared value of zero. Prediction is often used in signal encoding (for example, speech compression), because if the next values of a signal can be accurately predicted, then these samples need not be transmitted or stored. Prediction can also be used to reduce noise or interference and therefore enhance the signal quality if the adaptive processor is designed to only predict the signal and ignore random noise elements or known interference patterns.

As shown in Figure 1.27(c), another application of adaptive systems is system modeling of an unknown or difficult to characterize system. The desired signal (d) is the unknown system's output and the input to the unknown system and the adaptive processor (x) is a broadband test signal (perhaps white Gaussian noise). After adaptation, the

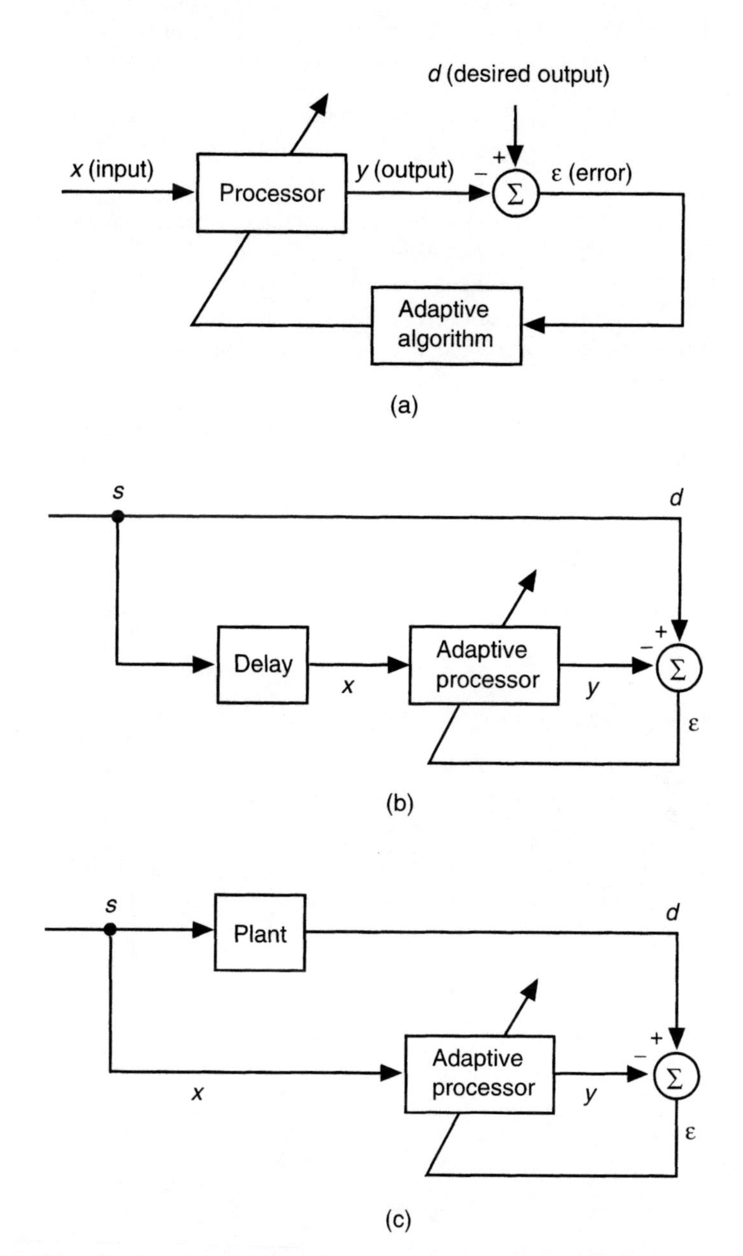

FIGURE 1.27 **(a)** Closed-loop adaptive system; (b) prediction; (c) system modeling.

unknown system is modeled by the final transfer function of the adaptive processor. By using an AR, MA, or ARMA adaptive processor, different system models can be obtained. The magnitude of the error (ε) can be used to judge the relative success of each model.

1.7.1 Wiener Filter Theory

The problem of determining the optimum linear filter given the structure shown in Figure 1.28 was solved by Norbert Wiener and others. The solution is referred to as the *Wiener filter.* The statement of the problem is as follows:

Determine a set of coefficients, w_k, that minimize the mean of the squared error of the filtered output as compared to some desired output. The error is written

$$e(n) = d(n) - \sum_{k=1}^{M} w_k^* u(n-k+1), \tag{1.106}$$

or in vector form

$$e(n) = d(n) - \mathbf{w}^H \mathbf{u}(n). \tag{1.107}$$

The mean squared error is a function of the tap weight vector **w** chosen and is written

$$J(\mathbf{w}) = E\{e(n)e^*(n)\}. \tag{1.108}$$

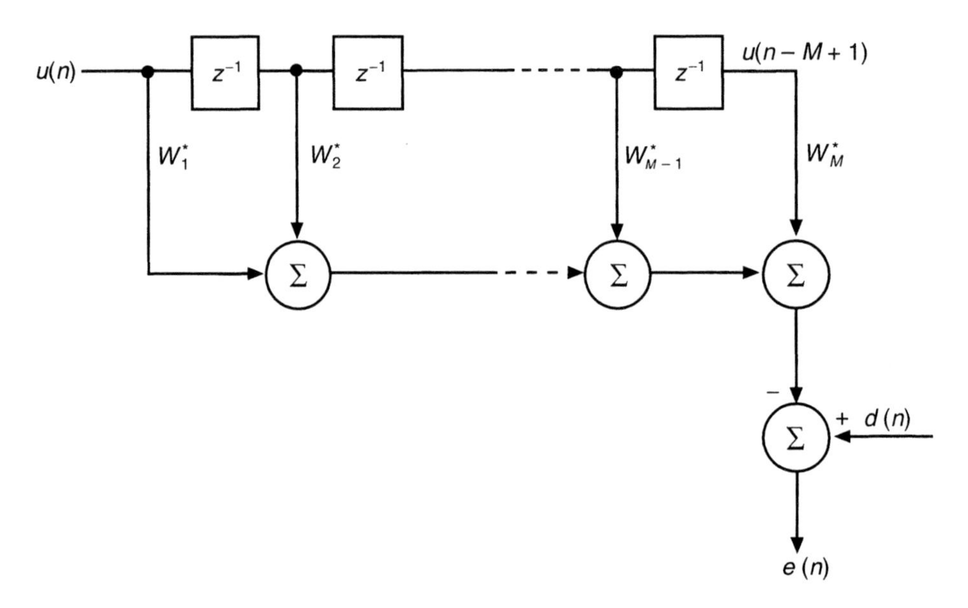

FIGURE 1.28 Wiener filter problem.

Substituting in the expression for $e(n)$ gives

$$J(\mathbf{w}) = E\Big\{d(n)d^*(n) - d(n)\mathbf{u}^H(n)\mathbf{w} - \mathbf{w}^H\mathbf{u}(n)d^*(n) + \mathbf{w}^H\mathbf{u}(n)\mathbf{u}^H(n)\mathbf{w}\Big\} \tag{1.109}$$

$$J(\mathbf{w}) = \text{var}\{d\} - \mathbf{p}^H\mathbf{w} - \mathbf{w}^H\mathbf{p} + \mathbf{w}^H\mathbf{R}\mathbf{w}, \tag{1.110}$$

where $\mathbf{p} = E\{\mathbf{u}(n)d^*(n)\}$, the vector that is the product of the cross correlation between the desired signal and each element of the input vector.

In order to minimize $J(\mathbf{w})$ with respect to $\mathbf{w}$, the tap weight vector, one must set the derivative of $J(\mathbf{w})$ with respect to $\mathbf{w}$ equal to zero. This will give an equation which, when solved for $\mathbf{w}$, gives $\mathbf{w}_0$, the optimum value of $\mathbf{w}$. Setting the total derivative equal to zero gives

$$-2\mathbf{p} + 2\mathbf{R}\mathbf{w}_0 = 0 \tag{1.111}$$

or

$$\mathbf{R}\mathbf{w}_0 = \mathbf{p}. \tag{1.112}$$

If the matrix $\mathbf{R}$ is invertible (nonsingular) then $\mathbf{w}_0$ can be solved as

$$\mathbf{w}_0 = \mathbf{R}^{-1}\mathbf{p}. \tag{1.113}$$

So the optimum tap weight vector depends on the autocorrelation of the input process and the cross correlation between the input process and the desired output. Equation (1.113) is called the *normal equation* because a filter derived from this equation will produce an error that is orthogonal (or normal) to each element of the input vector. This can be written

$$E\big\{\mathbf{u}(n)e_0^*(n)\big\} = 0. \tag{1.114}$$

It is helpful at this point to consider what must be known to solve the Wiener filter problem:

(1) The M x M autocorrelation matrix of $\mathbf{u}(n)$, the input vector

(2) The cross correlation vector between $\mathbf{u}(n)$ and $d(n)$ the desired response.

It is clear that knowledge of any individual $\mathbf{u}(n)$ will not be sufficient to calculate these statistics. One must take the ensemble average, $E\{\ \}$, to form both the autocorrelation and the cross correlation. In practice, a model is developed for the input process and from this model the second order statistics are derived.

A legitimate question at this point is: What is $d(n)$? It depends on the problem. One example of the use of Wiener filter theory is in linear predictive filtering. In this case, the desired signal is the next value of $\mathbf{u}(n)$, the input. The actual $\mathbf{u}(n)$ is always available one sample after the prediction is made and this gives the ideal check on the quality of the prediction.

1.7.2 LMS Algorithms

The LMS algorithm is the simplest and most used adaptive algorithm in use today. In this brief section, the LMS algorithm as it is applied to the adaptation of time-varying FIR filters (MA systems) and IIR filters (adaptive recursive filters or ARMA systems) is described. A detailed derivation, justification and convergence properties can be found in the references.

For the adaptive FIR system the transfer function is described by

$$y(n) = \sum_{q=0}^{Q-1} b_q(k)x(n-q), \tag{1.115}$$

where $b(k)$ indicates the time-varying coefficients of the filter. With an FIR filter the mean squared error performance surface in the multidimensional space of the filter coefficients is a quadratic function and has a single minimum mean squared error (MMSE). The coefficient values at the optimal solution is called the MMSE solution. The goal of the adaptive process is to adjust the filter coefficients in such a way that they move from their current position toward the MMSE solution. If the input signal changes with time, the adaptive system must continually adjust the coefficients to follow the MMSE solution. In practice, the MMSE solution is often never reached.

The LMS algorithm updates the filter coefficients based on the method of steepest descent. This can be described in vector notation as follows:

$$\mathbf{B}_{k+1} = \mathbf{B}_k - \mu \nabla_k \tag{1.116}$$

where $\mathbf{B}_k$ is the coefficient column vector, μ is a parameter that controls the rate of convergence and the gradient is approximated as

$$\nabla_k = \frac{\partial E\left[\epsilon_k^2\right]}{\partial \mathbf{B}_k} \cong -2\epsilon_k \mathbf{X}_k \tag{1.117}$$

where $\mathbf{X}_k$ is the input signal column vector and ε_k is the error signal as shown on Figure 1.27. Thus, the basic LMS algorithm can be written as

$$\mathbf{B}_{k+1} = \mathbf{B}_k + 2\mu\epsilon_k \mathbf{X}_k \tag{1.118}$$

The selection of the convergence parameter must be done carefully, because if it is too small the coefficient vector will adapt very slowly and may not react to changes in the input signal. If the convergence parameter is too large, the system will adapt to noise in the signal and may never converge to the MMSE solution.

For the adaptive IIR system the transfer function is described by

$$y(n) = \sum_{q=0}^{Q-1} b_q(k)x(n-q) - \sum_{p=1}^{P-1} a_p(k)y(n-p), \tag{1.119}$$

where $b(k)$ and $a(k)$ indicate the time-varying coefficients of the filter. With an IIR filter, the mean squared error performance surface in the multidimensional space of the filter coefficients is not a quadratic function and can have multiple minimums that may cause the adaptive algorithm to never reach the MMSE solution. Because the IIR system has poles, the system can become unstable if the poles ever move outside the unit circle during the adaptive process. These two potential problems are serious disadvantages of adaptive recursive filters that limit their application and complexity. For this reason, most applications are limited to a small number of poles. The LMS algorithm can again be used to update the filter coefficients based on the method of steepest descent. This can be described in vector notation as follows:

$$\mathbf{W}_{k+1} = \mathbf{W}_k - \mathbf{M}\nabla_k, \tag{1.120}$$

where $\mathbf{W}_k$ is the coefficient column vector containing the a and b coefficients, $\mathbf{M}$ is a diagonal matrix containing convergence parameters μ for the a coefficients and ν_0 through ν_{P-1} that controls the rate of convergence of the b coefficients. In this case, the gradient is approximated as

$$\nabla_k \cong -2\epsilon_k \left[\alpha_0 \ldots \alpha_{Q-1} \beta_1 \ldots \beta_P\right]^T, \tag{1.121}$$

where ε_k is the error signal as shown in Figure 1.27, and

$$\alpha_n(k) = x(k-n) + \sum_{q=0}^{Q-1} b_q(k)\alpha_n(k-q) \tag{1.122}$$

$$\beta_n(k) = y(k-n) + \sum_{p=0}^{P-1} b_q(k)\beta_n(k-p). \tag{1.123}$$

The selection of the convergence parameters must be done carefully because if they are too small the coefficient vector will adapt very slowly and may not react to changes in the input signal. If the convergence parameters are too large the system will adapt to noise in the signal or may become unstable. The proposed new location of the poles should also be tested before each update to determine if an unstable adaptive filter is about to be used. If an unstable pole location is found the update should not take place and the next update value may lead to a better solution.

1.8 REFERENCES

BRIGHAM, E. (1974). *The Fast Fourier Transform.* Englewood Cliffs, NJ: Prentice Hall.

CLARKSON, P. (1993). *Optimal and Adaptive Signal Processing.* FL: CRC Press.

ELIOTT, D.F. (Ed.). (1987). *Handbook of Digital Signal Processing.* San Diego, CA: Academic Press.

EMBREE, P. and KIMBLE, B. (1991). *C Language Algorithms for Digital Signal Processing.* Englewood Cliffs, NJ: Prentice Hall.

HARRIS, F. (1978). On the Use of Windows for Harmonic Analysis with the Discrete Fourier Transform. *Proceedings of the IEEE., 66,* (1), 51–83.

HAYKIN, S. (1986). *Adaptive Filter Theory.* Englewood Cliffs, NJ: Prentice Hall.

MCCLELLAN, J., PARKS, T. and RABINER, L.R. (1973). A Computer Program for Designing Optimum FIR Linear Phase Digital Filters. *IEEE Transactions on Audio and Electro-acoustics, AU-21.* (6), 506–526.

MOLER, C., LITTLE, J. and BANGERT, S. (1987). *PC-MATLAB User's Guide.* Sherbourne, MA: The Math Works.

OPPENHEIM, A. and SCHAFER, R. (1975). *Digital Signal Processing.* Englewood Cliffs, NJ: Prentice Hall.

OPPENHEIM, A. and SCHAFER, R. (1989). *Discrete-time Signal Processing.* Englewood Cliffs, NJ: Prentice Hall.

PAPOULIS, A. (1965). *Probability, Random Variables and Stochastic Processes.* New York: McGraw-Hill.

RABINER, L. and GOLD, B. (1975). *Theory and Application of Digital Signal Processing.* Englewood Cliffs, NJ: Prentice Hall.

STEARNS, S. and DAVID, R. (1988). *Signal Processing Algorithms.* Englewood Cliffs, NJ: Prentice Hall.

STEARNS, S. and DAVID, R. (1993). *Signal Processing Algorithms in FORTRAN and C.* Englewood Cliffs, NJ: Prentice Hall.

VAIDYANATHAN, P. (1993). *Multirate Systems and Filter Banks.* Englewood Cliffs, NJ: Prentice Hall.

WIDROW, B. and STEARNS, S. (1985). *Adaptive Signal Processing.* Englewood Cliffs, NJ: Prentice Hall.

CHAPTER 2

C PROGRAMMING FUNDAMENTALS

The purpose of this chapter is to provide the programmer with a complete overview of the fundamentals of the C programming language that are important in DSP applications. In particular, text manipulation, bitfields, enumerated data types, and unions are not discussed, because they have limited utility in the majority of DSP programs. Readers with C programming experience may wish to skip the bulk of this chapter with the possible exception of the more advanced concepts related to pointers and structures presented in sections 2.7 and 2.8. The proper use of pointers and data structures in C can make a DSP program easier to write and much easier for others to understand. Example DSP programs in this chapter and those which follow will clarify the importance of pointers and data structures in DSP programs.

2.1 THE ELEMENTS OF REAL-TIME DSP PROGRAMMING

The purpose of a *programming language* is to provide a tool so that a programmer can easily solve a problem involving the manipulation of some type of information. Based on this definition, the purpose of a DSP program is to manipulate a signal (a special kind of information) in such a way that the program solves a signal-processing problem. To do this, a DSP programming language must have five basic elements:

(1) A method of organizing different types of data (variables and data types)

(2) A method of describing the operations to be done (operators)

(3) A method of controlling the operations performed based on the results of operations (program control)

(4) A method of organizing the data and the operations so that a sequence of program steps can be executed from anywhere in the program (functions and data structures) and

(5) A method to move data back and forth between the outside world and the program (input/output)

These five elements are required for efficient programming of DSP algorithms. Their implementation in C is described in the remainder of this chapter.

As a preview of the C programming language, a simple real-time DSP program is shown in Listing 2.1. It illustrates each of the five elements of DSP programming. The listing is divided into six sections as indicated by the comments in the program. This simple DSP program gets a series of numbers from an input source such as an A/D converter (the function **getinput()** is not shown, since it would be hardware specific) and determines the average and variance of the numbers which were sampled. In signal-processing terms, the output of the program is the DC level and total AC power of the signal.

The first line of Listing 2.1, **main()**, declares that the program called *main,* which has no arguments, will be defined after the next left brace (**{** on the next line). The main program (called main because it is executed first and is responsible for the main control of the program) is declared in the same way as the functions. Between the left brace on the second line and the right brace half way down the page (before the line that starts **float average**...) are the statements that form the main program. As shown in this example, all statements in C end in a semicolon (**;**) and may be placed anywhere on the input line. In fact, all spaces and carriage control characters are ignored by most C compilers. Listing 2.1 is shown in a format intended to make it easier to follow and modify.

The third and fourth lines of Listing 2.1 are statements declaring the *functions* (**average**, **variance**, **sqrt**) that will be used in the rest of the main program (the function **sqrt()** is defined in the standard C library as discussed in the Appendix. This first section of Listing 2.1 relates to program organization (element four of the above list). The beginning of each section of the program is indicated by comments in the *program source code* (i. e., **/* section 1 */**). Most C compilers allow any sequence of characters (including multiple lines and, in some cases, nested comments) between the **/*** and ***/** delimiters.

Section two of the program declares the variables to be used. Some variables are declared as *single floating-point numbers* (such as **ave** and **var**); some variables are declared as *single integers* (such as **i**, **count**, and **number**); and some variables are *arrays* (such as **signal[100]**). This program section relates to element one, data organization.

Section three reads 100 floating-point values into an array called signal using a **for** *loop* (similar to a DO loop in FORTRAN). This loop is inside an infinite **while** *loop* that is common in real-time programs. For every 100 samples, the program will display the results and then get another 100 samples. Thus, the results are displayed in real-time. This section relates to element five (input/output) and element three (program control).

Section four of the example program uses the functions **average** and **variance**

```
main()                                     /* section 1 */
{
  float average(),variance(),sqrt();       /* declare functions */

  float signal[100],ave,var;               /*section 2 */
  int count,i;                             /* declare variables */

  while(1) {
    for(count = 0 ; count < 100 ; count++) { /* section 3 */
        signal[count] = getinput();        /* read input signal */
    }

    ave = average(signal,count);           /* section 4 */
    var = variance(signal,count);          /* calculate results */

    printf("\n\nAverage = %f",ave);        /* section 5 */
    printf(" Variance = %f",var);          /* print results */
  }
}
float average(float array[],int size)      /* section 6 */
{                                          /* function calculates average */
    int i;
    float sum = 0.0;                       /* intialize and declare sum */
    for(i = 0 ; i < size ; i++)
        sum = sum + array[i];              /* calculate sum */
    return(sum/size);                      /* return average */
}
float variance(float array[],int size)     /* function calculates variance */
{
    int i;                                 /* declare local variables */
    float ave;
    float sum = 0.0;                       /* intialize sum of signal */
    float sum2 = 0.0;                      /* sum of signal squared */
    for(i = 0 ; i < size ; i++) {
        sum = sum + array[i];
        sum2 = sum2 + array[i]*array[i];   /* calculate both sums */
    }
    ave = sum/size;                        /* calculate average */
    return((sum2 - sum*ave)/(size-1));     /* return variance */
}
```

Listing 2.1 Example C program that calculates the average and variance of a sequence of numbers.

to calculate the statistics to be printed. The variables **ave** and **var** are used to store the results and the library function **printf** is used to display the results. This part of the program relates to element four (functions and data structures) because the operations defined in functions **average** and **variance** are executed and stored.

Section five uses the library function **printf** to display the results **ave**, **var**, and also calls the function **sqrt** in order to display the standard deviation. This part of the program relates to element four (functions) and element five (input/output), because the operations defined in function **sqrt** are executed and the results are also displayed.

The two functions, **average** and **variance**, are defined in the remaining part of Listing 2.1. This last section relates primarily to element two (operators), since the detailed operation of each function is defined in the same way that the main program was defined. The function and argument types are defined and the local variables to be used in each function are declared. The operations required by each function are then defined followed by a return statement that passes the result back to the main program.

2.2 VARIABLES AND DATA TYPES

All programs work by manipulating some kind of information. A variable in C is defined by declaring that a sequence of characters (the variable identifier or name) are to be treated as a particular predefined type of data. An identifier may be any sequence of characters (usually with some length restrictions) that obeys the following three rules:

(1) All identifiers start with a letter or an underscore (_).

(2) The rest of the identifier can consist of letters, underscores, and/or digits.

(3) The rest of the identifier does not match any of the C keywords. (Check compiler implementation for a list of these.)

In particular, C is case sensitive; making the variables **Average**, **AVERAGE**, and **AVeRaGe** all different.

The C language supports several different data types that represent integers (declared **int**), floating-point numbers (declared **float** or **double**), and text data (declared **char**). Also, arrays of each variable type and pointers of each type may be declared. The first two types of numbers will be covered first followed by a brief introduction to arrays (covered in more detail with pointers in section 2.7). The special treatment of text using character arrays and strings will be discussed in Section 2.2.3.

2.2.1 Types of Numbers

A C program must declare the variable before it is used in the program. There are several types of numbers used depending on the format in which the numbers are stored (floating-point format or integer format) and the accuracy of the numbers (single-precision versus double-precision floating-point, for example). The following example program illustrates the use of five different types of numbers:

```
main()
{
    int i;              /* size dependent on implementation */
    short j;            /* 16 bit integer */
    long k;             /* 32 bit integer */
    float a;            /* single precision floating-point */
    double b;           /* double precision floating-point */
    k = 72000;
    j = k;
    i = k;
    b = 0.1;
    a = b;

    printf("\n%ld %d %d\n%20.15f\n%20.15f",k,j,i,b,a);
}
```

Three types of integer numbers (**int**, **short int**, and **long int**) and two types of floating-point numbers (**float** and **double**) are illustrated in this example. The actual sizes (in terms of the number of bytes used to store the variable) of these five types depends upon the implementation; all that is guaranteed is that a **short int** variable will not be larger than a **long int** and a **double** will be twice as large as a **float**. The size of a variable declared as just **int** depends on the compiler implementation. It is normally the size most conveniently manipulated by the target computer, thereby making programs using **int**s the most efficient on a particular machine. However, if the size of the integer representation is important in a program (as it often is) then declaring variables as **int** could make the program behave differently on different machines. For example, on a 16-bit machine, the above program would produce the following results:

```
72000 6464 6464
0.100000000000000
0.100000001490116
```

But on a 32-bit machine (using 32-bit **int**s), the output would be as follows:

```
72000 6464 72000
0.100000000000000
0.100000001490116
```

Note that in both cases the **short** and **long** variables, **k** and **j**, (the first two numbers displayed) are the same, while the third number, indicating the **int i**, differs. In both cases, the value 6464 is obtained by masking the lower 16 bits of the 32-bit **k** value. Also, in both cases, the floating-point representation of 0.1 with 32 bits (**float**) is accurate to eight decimal places (seven places is typical). With 64 bits it is accurate to at least 15 places.

Thus, to make a program truly portable, the program should contain only **short int** and **long int** declarations (these may be abbreviated **short** and **long**). In addition to the five types illustrated above, the three **int**s can be declared as unsigned by preceding the declaration with unsigned. Also, as will be discussed in more detail in the next section concerning text data, a variable may be declared to be only one byte long by declaring it a **char** (**signed** or **unsigned**). The following table gives the typical sizes and ranges of the different variable types for a 32-bit machine (such as a VAX) and a 16-bit machine (such as the IBM PC).

Variable Declaration	16-bit Machine Size (bits)	16-bit Machine Range	32-bit Machine Size (bits)	32-bit Machine Range
char	8	−128 to 127	8	−128 to 127
unsigned char	8	0 to 255	8	0 to 255
int	16	−32768 to 32767	32	±2.1e9
unsigned int	16	0 to 65535	32	0 to 4.3e9
short	16	−32768 to 32767	16	−32768 to 32767
unsigned short	16	0 to 65535	16	0 to 65535
long	32	±2.1e9	32	±2.1e9
unsigned long	32	0 to 4.3e9	32	0 to 4.3e9
float	32	±1.0e±38	32	±1e±38
double	64	±1.0e±306	64	±1e±308

2.2.2 Arrays

Almost all high-level languages allow the definition of indexed lists of a given data type, commonly referred to as *arrays.* In C, all data types can be declared as an array simply by placing the number of elements to be assigned to the array in brackets after the array name. *Multidimensional arrays* can be defined simply by appending more brackets containing the array size in each dimension. Any N-dimensional array is defined as follows:

```
type name[size1][size2] ... [sizeN];
```

For example, each of the following statements are valid array definitions:

```
unsigned int list[10];
double input[5];
short int x[2000];
char input_buffer[20];
unsigned char image[256][256];
int matrix[4][3][2];
```

Note that the array definition **unsigned char image[256][256]** could define an 8-bit, 256 by 256 image plane where a grey scale image is represented by values from 0 to 255. The last definition defines a three-dimensional matrix in a similar fashion. One difference between C and other languages is that arrays are referenced using brackets to enclose each index. Thus, the image array, as defined above, would be referenced as **image[i][j]** where **i** and **j** are row and column indices, respectively. Also, the first element in all array indices is zero and the last element is *N–1*, where *N* is the size of the array in a particular dimension. Thus, an assignment of the first element of the five element, one-dimensional array **input** (as defined above) such as **input[0]=1.3;** is legal while **input[5]=1.3;** is not.

Arrays may be *initialized* when they are declared. The values to initialize the array are enclosed in one or more sets of braces (**{}**) and the values are separated by commas. For example, a one-dimensional array called vector can be declared and initialized as follows:

```
int vector[6] = { 1, 2, 3, 5, 8, 13 };
```

A two-dimensional array of six double-precision floating-point numbers can be declared and initialized using the following statement:

```
double a[3][2] = {
              { 1.5, 2.5 },
              { 1.1e-5 , 1.7e5 },
              { 1.765 , 12.678 }
                      };
```

Note that commas separate the three sets of inner braces that designate each of the three rows of the matrix **a**, and that each array initialization is a statement that must end in a semicolon.

2.3 OPERATORS

Once variables are defined to be a given size and type, some sort of manipulation must be performed using the variables. This is done by using *operators*. The C language has more operators than most languages; in addition to the usual assignment and arithmetic operators, C also has bitwise operators and a full set of logical operators. Some of these operators (such as bitwise operators) are especially important in order to write DSP programs that utilize the target processor efficiently.

2.3.1 Assignment Operators

The most basic operator is the *assignment operator* which, in C, is the single equal sign (**=**). The value on the right of the equal sign is assigned to the variable on the left. Assignment statements can also be stacked, as in the statement **a=b=1;**. In this case, the statement is evaluated right to left so that 1 is assigned to **b** and **b** is assigned to **a**. In C,

a=ave(x) is an expression, while **a=ave(x);** is a statement. The addition of the semicolon tells the compiler that this is all that will be done with the result from the function **ave(x)**. An expression always has a value that can be used in other expressions. Thus, **a=b+(c=ave(x));** is a legal statement. The result of this statement would be that the result returned by **ave(x)** is assigned to **c** and **b+c** is assigned to **a**. C also allows multiple expressions to be placed within one statement by separating them with the commas. Each expression is evaluated left to right, and the entire expression (comprised of more than one expression) assumes the value of the last expression which is evaluated. For example, **a=(olda=a,ave(x));** assigns the current value of **a** to **olda**, calls the function **ave(x)** and then assigns the value returned by **ave(x)** to **a**.

2.3.2 Arithmetic and Bitwise Operators

The usual set of binary arithmetic operators (operators which perform arithmetic on two operands) are supported in C using the following symbols:

*	multiplication
/	division
+	addition
-	subtraction
%	modulus (integer remainder after division)

The first four operators listed are defined for all types of variables (**char**, **int**, **float**, and **double**). The modulus operator is only defined for integer operands. Also, there is no exponent operator in C; this floating-point operation is supported using a *simple function call* (see the Appendix for a description of the **pow** function).

In C, there are three unary arithmetic operators which require only one operand. First is the unary minus operator (for example, **-i**, where **i** is an **int**) that performs a two's-complement change of sign of the integer operand. The unary minus is often useful when the exact hardware implementation of a digital-signal processing algorithm must be simulated. The other two unary arithmetic operators are increment and decrement, represented by the symbols **++** and **--**, respectively. These operators add or subtract one from any integer variable or pointer. The operand is often used in the middle of an expression, and the increment or decrement can be done before or after the variable is used in the expression (depending on whether the operator is before or after the variable). Although the use of **++** and **--** is often associated with pointers (see section 2.7), the following example illustrates these two powerful operators with the ints **i**, **j**, and **k**:

```
i = 4;
j = 7;
k = i++ + j;          /* i is incremented to 5, k = 11 */
k = k + --j;          /* j is decremented to 6, k = 17 */
k = k + i++;          /* i is incremented to 6, k = 22 */
```

Binary bitwise operations are performed on integer operands using the following symbols:

&	bitwise AND
|	bitwise OR
^	bitwise exclusive OR
<<	arithmetic shift left (number of bits is operand)
>>	arithmetic shift right (number of bits is operand)

The unary bitwise NOT operator, which inverts all the bits in the operand, is implemented with the **~** symbol. For example, if **`i`** is declared as an **`unsigned int`**, then **`i = ~0;`** sets **`i`** to the maximum integer value for an **`unsigned int`**.

2.3.3 Combined Operators

C allows operators to be combined with the assignment operator (**=**) so that almost any statement of the form

```
<variable> = <variable> <operator> <expression>
```

can be replaced with

```
<variable> <operator> = <expression>
```

where **`<variable>`** represents the same variable name in all cases. For example, the following pairs of expressions involving **`x`** and **`y`** perform the same function:

```
x = x + y;          x += y;
x = x - y;          x -= y;
x = x * y;          x *= y;
x = x / y;          x /= y;
x = x % y;          x %= y;
x = x & y;          x &= y;
x = x : y;          x := y;
x = x ^ y;          x ^= y;
x = x << y;         x <<= y;
x = x >> y;         x >>= y;
```

In many cases, the left-hand column of statements will result in a more readable and easier to understand program. For this reason, use of combined operators is often avoided. Unfortunately, some compiler implementations may generate more efficient code if the combined operator is used.

2.3.4 Logical Operators

Like all C expressions, an expression involving a logical operator also has a value. A logical operator is any operator that gives a result of true or false. This could be a comparison between two values, or the result of a series of ANDs and ORs. If the result of a logical operation is true, it has a nonzero value; if it is false, it has the value 0. Loops and if statements (covered in section 2.4) check the result of logical operations and change program flow accordingly. The nine logical operators are as follows:

`<` less than

`<=` less than or equal to

`==` equal to

`>=` greater than or equal to

`>` greater than

`!=` not equal to

`&&` logical AND

`||` logical OR

`!` logical NOT (unary operator)

Note that `==` can easily be confused with the assignment operator (`=`) and will result in a valid expression because the assignment also has a value, which is then interpreted as true or false. Also, `&&` and `||` should not be confused with their bitwise counterparts (`&` and `|`) as this may result in hard to find logic problems, because the bitwise results may not give true or false when expected.

2.3.5 Operator Precedence and Type Conversion

Like all computer languages, C has an operator precedence that defines which operators in an expression are evaluated first. If this order is not desired, then parentheses can be used to change the order. Thus, things in parentheses are evaluated first and items of equal precedence are evaluated from left to right. The operators contained in the parentheses or expression are evaluated in the following order (listed by decreasing precedence):

`++,--` increment, decrement

`-` unary minus

`*,/,%` multiplication, division, modulus

`+,-` addition, subtraction

`<<,>>` shift left, shift right

`<,<=,>=,>` relational with less than or greater than

`==,!=` equal, not equal

`&` bitwise AND

`^` bitwise exclusive OR

`|` bitwise OR

`&&` logical AND

`||` logical OR

Statements and expressions using the operators just described should normally use variables and constants of the same type. If, however, you mix types, C doesn't stop dead (like Pascal) or produce a strange unexpected result (like FORTRAN). Instead, C uses a set of rules to make type conversions automatically. The two basic rules are:

(1) If an operation involves two types, the value with a lower rank is converted to the type of higher rank. This process is called promotion and the ranking from highest to lowest type is double, float, long, int, short, and char. Unsigned of each of the types outranks the individual signed type.

(2) In an assignment statement, the final result is converted to the type of the variable that is being assigned. This may result in promotion or demotion where the value is truncated to a lower ranking type.

Usually these rules work quite well, but sometimes the conversions must be stated explicitly in order to demand that a conversion be done in a certain way. This is accomplished by *type casting* the quantity by placing the name of the desired type in parentheses before the variable or expression. Thus, if **`i`** is an **`int`**, then the statement **`i=10*(1.55+1.67);`** would set **`i`** to 32 (the truncation of 32.2), while the statement **`i=10*((int)1.55+1.67);`** would set **`i`** to 26 (the truncation of 26.7 since **`(int)1.55`** is truncated to 1).

2.4 PROGRAM CONTROL

The large set of operators in C allows a great deal of programming flexibility for DSP applications. Programs that must perform fast binary or logical operations can do so without using special functions to do the bitwise operations. C also has a complete set of program control features that allow conditional execution or repetition of statements based on the result of an expression. Proper use of these control structures is discussed in section 2.11.2, where structured programming techniques are considered.

2.4.1 Conditional Execution: `if-else`

In C, as in many languages, the **`if`** statement is used to conditionally execute a series of statements based on the result of an expression. The **`if`** statement has the following generic format:

```
if(value)
    statement1;
else
    statement2;
```

where **`value`** is any expression that results in (or can be converted to) an integer value. If value is nonzero (indicating a true result), then **`statement1`** is executed; otherwise, **`statement2`** is executed. Note that the result of an expression used for **`value`** need not be the result of a logical operation—all that is required is that the expression results in a zero value when **`statement2`** should be executed instead of **`statement1`**. Also, the

else statement2; portion of the above form is optional, allowing **statement1** to be skipped if value is false.

When more than one statement needs to be executed if a particular value is true, a compound statement is used. A *compound statement* consists of a left brace (**{**), some number of statements (each ending with a semicolon), and a right brace (**}**). Note that the body of the **main()** program and functions in Listing 2.1 are compound statements. In fact, a single statement can be replaced by a compound statement in any of the control structures described in this section. By using compound statements, the **if-else** control structure can be nested as in the following example, which converts a floating-point number (**result**) to a 2-bit twos complement number (**out**):

```
if(result > 0) {            /* positive outputs */
    if(result > sigma)
        out = 1;            /* biggest output */
    else
        out = 0;            /* 0 < result <= sigma */
}
else {                      /* negative outputs */
    if(result < sigma)
        out = 2;            /* smallest output */
    else
        out = 1;            /* sigma <= result <= 0 */
}
```

Note that the inner **if-else** statements are compound statements (each consisting of two statements), which make the braces necessary in the outer **if-else** control structure (without the braces there would be too many **else** statements, resulting in a compilation error).

2.4.2 The switch Statement

When a program must choose between several alternatives, the **if-else** statement becomes inconvenient and sometimes inefficient. When more than four alternatives from a single expression are chosen, the **switch** statement is very useful. The basic form of the **switch** statement is as follows:

```
switch(integer expression) {
    case constant1:
        statements;      (optional)
        break;           (optional)
    case constant2:
        statements;      (optional)
        break;           (optional)
        . . . . .        (more optional statements)
    default:             (optional)
        statements;      (optional)
}
```

Program control jumps to the statement after the case label with the constant (an integer or single character in quotes) that matches the result of the integer expression in the **switch** statement. If no constant matches the expression value, control goes to the statement following the default label. If the default label is not present and no matching case labels are found, then control proceeds with the next statement following the **switch** statement. When a matching constant is found, the remaining statements after the corresponding case label are executed until the end of the switch statement is reached, or a **break** statement is reached that redirects control to the next statement after the **switch** statement. A simple example is as follows:

```
switch(i) {
    case 0:
        printf("\nError: I is zero");
        break;
    case 1:
        j = k*k;
        break;
    default:
        j = k*k/i;
}
```

The use of the **break** statement after the first two case statements is required in order to prevent the next statements from being executed (a **break** is not required after the last **case** or **default** statement). Thus, the above code segment sets **j** equal to **k*k/i**, unless **i** is zero, in which case it will indicate an error and leave **j** unchanged. Note that since the divide operation usually takes more time than the case statement branch, some execution time will be saved whenever **i** equals 1.

2.4.3 Single-Line Conditional Expressions

C offers a way to express one **if-else** control structure in a single line. It is called a *conditional expression,* because it uses the conditional operator, **?:**, which is the only trinary operator in C. The general form of the conditional expression is:

```
expression1 ? expression2 : expression3
```

If **expression1** is true (nonzero), then the whole conditional expression has the value of **expression2**. If **expression1** is false (0), the whole expression has the value of **expression3**. One simple example is finding the maximum of two expressions:

```
maxdif = (a0 > a2) ? a0-a1 : a2-a1;
```

Conditional expressions are not necessary, since **if-else** statements can provide the same function. Conditional expressions are more compact and sometimes lead to more efficient machine code. On the other hand, they are often more confusing than the familiar **if-else** control structure.

2.4.4 Loops: while, do-while, and for

C has three control structures that allow a statement or group of statements to be repeated a fixed or variable number of times. The **while** loop repeats the statements until a test expression becomes false, or zero. The decision to go through the loop is made before the loop is ever started. Thus, it is possible that the loop is never traversed. The general form is:

```
while(expression)
    statement
```

where **statement** can be a single statement or a compound statement enclosed in braces. An example of the latter that counts the number of spaces in a null-terminated string (an array of characters) follows:

```
space_count = 0;        /* space_count is an int */
i = 0;                  /* array index, i = 0 */
while(string[i]) {
    if(string[i] == ' ') space_count++;.
    i++;                /* next char */
}
```

Note that if the string is zero length, then the value of **string[i]** will initially point to the null terminator (which has a zero or false value) and the **while** loop will not be executed. Normally, the **while** loop will continue counting the spaces in the string until the null terminator is reached.

The **do-while** loop is used when a group of statements need to be repeated and the exit condition should be tested at the end of the loop. The decision to go through the loop one more time is made after the loop is traversed so that the loop is always executed at least once. The format of **do-while** is similar to the **while** loop, except that the **do** keyword starts the statement and **while(expression)** ends the statement. A single or compound statement may appear between the do and the while keywords. A common use for this loop is in testing the bounds on an input variable as the following example illustrates:

```
do {
    printf("\nEnter FFT length (less than 1025) :");
    scanf("%d",&fft_length);
} while(fft_length > 1024);
```

In this code segment, if the integer **fft_length** entered by the user is larger than 1024, the user is prompted again until the **fft_length** entered is 1024 or less.

The **for** loop combines an initialization statement, an end condition statement, and an action statement (executed at the end of the loop) into one very powerful control structure. The standard form is:

```
for(initialize ; test condition ; end update)
    statement;
```

The three expressions are all optional (**for(;;);** is an infinite loop) and the statement may be a single statement, a compound statement or just a semicolon (a null statement). The most frequent use of the **for** loop is indexing an array through its elements. For example,

```
for(i = 0 ; i < length ; i++) a[i] = 0;
```

sets the elements of the array **a** to zero from **a[0]** up to and including **a[length-1]**. This **for** statement sets **i** to zero, checks to see if **i** is less than **length**, if so it executes the statement **a[i]=0;**, increments **i**, and then repeats the loop until **i** is equal to **length**. The integer **i** is incremented or updated at the end of the loop and then the test condition statement is executed. Thus, the statement after a **for** loop is only executed if the test condition in the **for** loop is true. **For** loops can be much more complicated, because each statement can be multiple expressions as the following example illustrates:

```
for(i = 0 , i3 = 1 ; i < 25 ; i++ , i3 = 3*i3)
    printf("\n%d %d",i,i3);
```

This statement uses two **int**s in the **for** loop (**i**, **i3**) to print the first 25 powers of 3. Note that the end condition is still a single expression (**i < 25**), but that the initialization and end expressions are two assignments for the two integers separated by a comma.

2.4.5 Program Jumps: break, continue, and goto

The loop control structures just discussed and the conditional statements (**if**, **if-else**, and **switch**) are the most important control structures in C. They should be used exclusively in the majority of programs. The last three control statements (**break**, **continue**, and **goto**) allow for conditional program jumps. If used excessively, they will make a program harder to follow, more difficult to debug, and harder to modify.

The **break** statement, which was already illustrated in conjunction with the switch statement, causes the program flow to break free of the **switch**, **for**, **while**, or **do-while** that encloses it and proceed to the next statement after the associated control structure. Sometimes **break** is used to leave a loop when there are two or more reasons to end the loop. Usually, however, it is much clearer to combine the end conditions in a single logical expression in the loop test condition. The exception to this is when a large number of executable statements are contained in the loop and the result of some statement should cause a premature end of the loop (for example, an end of file or other error condition).

The **continue** statement is almost the opposite of **break**; the **continue** causes the rest of an iteration to be skipped and the next iteration to be started. The **continue** statement can be used with **for**, **while**, and **do-while** loops, but cannot be used with **switch**. The flow of the loop in which the **continue** statement appears is interrupted, but the loop is not terminated. Although the **continue** statement can result in very hard-to-follow code, it can shorten programs with nested **if-else** statements inside one of three loop structures.

The **goto** statement is available in C, even though it is never required in C programming. Most programmers with a background in FORTRAN or BASIC computer languages (both of which require the **goto** for program control) have developed bad programming habits that make them depend on the **goto**. The **goto** statement in C uses a label rather than a number making things a little better. For example, one possible legitimate use of **goto** is for consolidated error detection and cleanup as the following simple example illustrates:

```
.
.
program statements
.
.
status = function_one(alpha,beta,constant);
if(status != 0) goto error_exit;
.
.
more program statements
.
.
status = function_two(delta,time);
if(status != 0) goto error_exit;
.
.
.
error_exit:          /*end up here from all errors */
    switch(status) {
        case 1:
            printf("\nDivide by zero error\n");
            exit();
        case 2:
            printf("\nOut of memory error\n");
            exit();
        case 3:
            printf("\nLog overflow error\n");
            exit();
        default:
            printf("\nUnknown error\n");
            exit();
    }
```

In the above example, both of the fictitious functions, **function_one** and **function_two** (see the next section concerning the definition and use of functions), perform some set of operations that can result in one of several errors. If no errors are detected, the function returns zero and the program proceeds normally. If an error is detected, the integer **status** is set to an error code and the program jumps to the label **error_exit** where a message indicating the type of error is printed before the program is terminated.

2.5 FUNCTIONS

All C programs consist of one or more functions. Even the program executed first is a function called **main()**, as illustrated in Listing 2.1. Thus, unlike other programming languages, there is no distinction between the main program and programs that are called by the main program (sometimes called *subroutines*). A C function may or may not return a value thereby removing another distinction between subroutines and functions in languages such as FORTRAN. Each C function is a program equal to every other function. Any function can call any other function (a function can even call itself), or be called by any other function. This makes C functions somewhat different than Pascal procedures, where procedures nested inside one procedure are ignorant of procedures elsewhere in the program. It should also be pointed out that unlike FORTRAN and several other languages, C always passes functions arguments by value not by reference. Because arguments are passed by value, when a function must modify a variable in the calling program, the C programmer must specify the function argument as a pointer to the beginning of the variable in the calling program's memory (see section 2.7 for a discussion of pointers).

2.5.1 Defining and Declaring Functions

A function is defined by the function type, a function name, a pair of parentheses containing an optional formal argument list, and a pair of braces containing the optional executable statements. The general format for ANSI C is as follows:

```
type name(formal argument list with declarations)
{
    function body
}
```

The **type** determines the type of value the function returns, not the type of arguments. If no **type** is given, the function is assumed to return an **int** (actually, a variable is also assumed to be of type **int** if no type specifier is provided). If a function does not return a value, it should be declared with the type **void**. For example, Listing 2.1 contains the function average as follows:

```
float average(float array[],int size)
{
    int i;
    float sum = 0.0;      /* initialize and declare sum */
    for(i = 0 ; i < size ; i++)
        sum = sum + array[i];    /* calculate sum */
    return(sum/size);     /* return average */
}
```

The first line in the above code segment declares a function called **average** will return a single-precision floating-point value and will accept two arguments. The two *argument names* (**array** and **size**) are defined in the formal *argument list* (also called the formal parameter list). The type of the two arguments specify that **array** is a one-dimensional array (of unknown length) and **size** is an **int**. Most modern C compilers allow the argument declarations for a function to be condensed into the argument list.

Note that the variable **array** is actually just a pointer to the beginning of the **float** array that was allocated by the calling program. By passing the pointer, only *one value* is passed to the function and not the large floating-point array. In fact, the function could also be declared as follows:

```
float average(float *array,int size)
```

This method, although more correct in the sense that it conveys what is passed to the function, may be more confusing because the function body references the variable as **array[i]**.

The body of the function that defines the executable statements and local variables to be used by the function are contained between the two braces. Before the ending brace (**}**), a return statement is used to return the **float** result back to the calling program. If the function did not return a value (in which case it should be declared **void**), simply omitting the return statement would return control to the calling program after the last statement before the ending brace. When a function with no return value must be terminated before the ending brace (if an error is detected, for example), a **return;** statement without a value should be used. The parentheses following the return statement are only required when the result of an expression is returned. Otherwise, a constant or variable may be returned without enclosing it in parentheses (for example, **return 0;** or **return n;**).

Arguments are used to convey values from the calling program to the function. Because the arguments are passed by value, a local copy of each argument is made for the function to use (usually the variables are stored on the stack by the calling program). The local copy of the arguments may be freely modified by the function body, but will not change the values in the calling program since only the copy is changed. The return statement can communicate one value from the function to the calling program. Other than this returned value, the function may not directly communicate back to the calling program. This method of passing arguments by value, such that the calling program's variables are isolated from the function, avoids the common problem in FORTRAN where modifications of arguments by a function get passed back to the calling program, resulting in the occasional modification of constants within the calling program.

When a function must return more than one value, one or more pointer arguments must be used. The calling program must allocate the storage for the result and pass the function a pointer to the memory area to be modified. The function then gets a copy of the pointer, which it uses (with the indirection operator, **`*`**, discussed in more detail in Section 2.7.1) to modify the variable allocated by the calling program. For example, the functions **`average`** and **`variance`** in Listing 2.1 can be combined into one function that passes the arguments back to the calling program in two **`float`** pointers called **`ave`** and **`var`**, as follows:

```
void stats(float *array,int size,float *ave,float *var)
{
    int i;
    float sum = 0.0;        /* initialize sum of signal */
    float sum2 = 0.0;       /* sum of signal squared */
    for(i = 0 ; i < size ; i++) {
        sum = sum + array[i];      /* calculate sums */
        sum2 = sum2 + array[i]*array[i];
}
    *ave = sum/size;        /* pass average and variance */
    *var = (sum2-sum*(*ave))/(size-1);
}
```

In this function, no value is returned, so it is declared type **`void`** and no return statement is used. This **`stats`** function is more efficient than the functions **`average`** and **`variance`** together, because the sum of the array elements was calculated by both the average function and the variance function. If the variance is not required by the calling program, then the average function alone is much more efficient, because the sum of the squares of the array elements is not required to determine the average alone.

2.5.2 Storage Class, Privacy, and Scope

In addition to type, variables and functions have a property called *storage class*. There are four storage classes with four storage class designators: **`auto`** for *automatic variables* stored on the stack, **`extern`** for *external variables* stored outside the current module, **`static`** for variables known *only in the current module,* and **`register`** for *temporary variables* to be stored in one of the registers of the target computer. Each of these four storage classes defines the scope or degree of the privacy a particular variable or function holds. The storage class designator keyword (**`auto`**, **`extern`**, **`static`**, or **`register`**) must appear first in the variable declaration before any type specification. The privacy of a variable or function is the degree to which other modules or functions cannot access a variable or call a function. Scope is, in some ways, the complement of privacy because the scope of a variable describes how many modules or functions have access to the variable.

Auto variables can only be declared within a function, are created when the function is invoked, and are lost when the function is exited. **Auto** variables are known only to the function in which they are declared and do not retain their value from one invocation of a function to another. Because **auto** variables are stored on a stack, a function that uses only **auto** variables can call itself recursively. The **auto** keyword is rarely used in C programs, since variables declared within functions default to the **auto** storage class.

Another important distinction of the **auto** storage class is that an **auto** variable is only defined within the control structure that surrounds it. That is, the scope of an **auto** variable is limited to the expressions between the braces (**{** and **}**) containing the variable declaration. For example, the following simple program would generate a compiler error, since **j** is unknown outside of the **for** loop:

```
main()
{
    int i;
    for(i = 0 ; i < 10 ; i++) {
        int j;          /* declare j here */
        j = i*i;
        printf("%d",j);
    }
    printf("%d",j);     /* j unknown here */
}
```

Register variables have the same scope as **auto** variables, but are stored in some type of register in the target computer. If the target computer does not have registers, or if no more registers are available in the target computer, a variable declared as **register** will revert to **auto**. Because almost all microprocessors have a large number of registers that can be accessed much faster than outside memory, **register** variables can be used to speed up program execution significantly. Most compilers limit the use of **register** variables to pointers, integers, and characters, because the target machines rarely have the ability to use registers for floating-point or double-precision operations.

Extern variables have the broadest scope. They are known to all functions in a module and are even known outside of the module in that they are declared. **Extern** variables are stored in their own separate data area and must be declared outside of any functions. Functions that access **extern** variables must be careful not to call themselves or call other functions that access the same **extern** variables, since **extern** variables retain their values as functions are entered and exited. **Extern** is the default storage class for variables declared outside of functions and for the functions themselves. Thus, functions not declared otherwise may be invoked by any function in a module as well as by functions in other modules.

Static variables differ from **extern** variables only in scope. A **static** variable declared outside of a function in one module is known only to the functions in that module. A **static** variable declared inside a function is known only to the function in which it is declared. Unlike an **auto** variable, a **static** variable retains its value from one invocation of a function to the next. Thus, **static** refers to the memory area assigned to the variable and does not indicate that the value of the variable cannot be changed. Functions may also be declared **static**, in which case the function is only known to other functions in the same module. In this way, the programmer can prevent other modules (and, thereby, other users of the object module) from invoking a particular function.

2.5.3 Function Prototypes

Although not in the original definition of the C language, *function prototypes,* in one form or another, have become a standard C compiler feature. A function prototype is a statement (which must end with a semicolon) describing a particular function. It tells the compiler the type of the function (that is, the type of the variable it will return) and the type of each argument in the formal argument list. The function named in the function prototype may or may not be contained in the module where it is used. If the function is not defined in the module containing the prototype, the prototype must be declared external. All C compilers provide a series of header files that contain the function prototypes for all of the standard C functions. For example, the prototype for the **stats** function defined in Section 2.5.1 is as follows:

```
extern void stats(float *,int,float *,float *);
```

This prototype indicates that **stats** (which is assumed to be in another module) returns no value and takes four arguments. The first argument is a pointer to a **float** (in this case, the array to do statsistics on). The second argument is an integer (in this case, giving the size of the array) and the last two arguments are pointers to **floats** which will return the average and variance results.

The result of using function prototypes for all functions used by a program is that the compiler now knows what type of arguments are expected by each function. This information can be used in different ways. Some compilers convert whatever type of actual argument is used by the calling program to the type specified in the function prototype and issue a warning that a data conversion has taken place. Other compilers simply issue a warning indicating that the argument types do not agree and assume that the programmer will fix it if such a mismatch is a problem. The ANSI C method of declaring functions also allows the use of a dummy variable with each formal parameter. In fact, when this ANSI C approach is used with dummy arguments, the only difference between function prototypes and function declarations is the semicolon at the end of the function prototype and the possible use of **extern** to indicate that the function is defined in another module.

2.6 MACROS AND THE C PREPROCESSOR

The C *preprocessor* is one of the most useful features of the C programming language. Although most languages allow compiler constants to be defined and used for conditional compilation, few languages (except for assembly language) allow the user to define macros. Perhaps this is why C is occasionally referred to as a *portable macro assembly language.* The large set of preprocessor directives can be used to completely change the look of a C program such that it is very difficult for anyone to decipher. On the other hand, the C preprocessor can be used to make complicated programs easy to follow, very efficient, and easy to code. The remainder of this chapter and the programs discussed in this book hopefully will serve to illustrate the latter advantages of the C preprocessor.

The C preprocessor allows *conditional compilation* of program segments, *user-defined symbolic replacement* of any text in the program (called *aliases* as discussed in Section 2.6.2), and *user-defined multiple parameter macros.* All of the preprocessor directives are evaluated before any C code is compiled and the directives themselves are removed from the program before compilation begins. Each preprocessor directive begins with a pound sign (**#**) followed by the preprocessor keyword. The following list indicates the basic use of each of the most commonly used preprocessor directives:

`#define NAME macro`	Associate symbol **NAME** with **macro** definition (optional parameters)
`#include "file"`	Copy named **file** (with directory specified) into current compilation
`#include <file>`	Include **file** from standard C library
`#if expression`	Conditionally compile the following code if result of **expression** is true
`#ifdef symbol`	Conditionally compile the following code if the **symbol** is defined
`#ifndef symbol`	Conditionally compile the following code if the **symbol** is not defined
`#else`	Conditionally compile the following code if the associated **#if** is not true
`#endif`	Indicates the end of previous **#else**, **#if**, **#ifdef**, or **#ifndef**
`#undef macro`	Undefine previously defined **macro**

2.6.1 Conditional Preprocessor Directives

Most of the above preprocessor directives are used for conditional compilation of portions of a program. For example, in the following version of the **stats** function (described previously in section 2.5.1), the definition of DEBUG is used to indicate that the print statements should be compiled:

```
void stats(float *array,int size,float *ave,float *var)
{
    int i;
    float sum = 0.0;        /* initialize sum of signal */
    float sum2 = 0.0;       /* sum of signal squared */
    for(i = 0 ; i < size ; 1++) {
        sum = sum + array[i];
        sum2 = sum2 + array[i]*array][i];     /* calculate sums */
}

#ifdef DEBUG
    printf("\nIn stats sum = %f sum2 = %f",sum,sum2);
    printf("\nNumber of array elements = %d",size);
#endif

    *ave = sum/size;        /* pass average */
    *var = (sum2 - sum* (*ave))/(size-1);     /* pass variance */
}
```

If the preprocessor parameter DEBUG is defined anywhere before the **#ifdef DEBUG** statement, then the **printf** statements will be compiled as part of the program to aid in debugging **stats** (or perhaps even the calling program). Many compilers allow the definition of preprocessor directives when the compiler is invoked. This allows the DEBUG option to be used with no changes to the program text.

2.6.2 Aliases and Macros

Of all the preprocessor directives, the **#define** directive is the most powerful because it allows aliases and multiple parameter macros to be defined in a relatively simple way. The most common use of **#define** is a macro with no arguments that replaces one string (the macro name) with another string (the macro definition). In this way, an alias can be given to any string including all of the C keywords. For example:

```
#define DO for(
```

replaces every occurrence of the string **DO** (all capital letters so that it is not confused with the C keyword **do**) with the four-character string **for(**. Similarly, new aliases of all the C keywords could be created with several **#define** statements (although this seems silly since the C keywords seem good enough). Even single characters can be aliased. For example, **BEGIN** could be aliased to **{** and **END** could be aliased to **}**, which makes a C program look more like Pascal.

The **#define** directive is much more powerful when parameters are used to create a true macro. The above **DO** macro can be expanded to define a simple FORTRAN style DO loop as follows:

```
#define DO(var,beg,end) for(var=beg; var<=end; var++)
```

The three macro parameters **var**, **beg**, and **end** are the variable, the beginning value, and the ending value of the DO loop. In each case, the macro is invoked and the string placed in each argument is used to expand the macro. For example,

```
DO(i,1,10)
```

expands to

```
for(i=1; i<=10; i++)
```

which is the valid beginning of a **for** loop that will start the variable **i** at 1 and stop it at 10. Although this DO macro does shorten the amount of typing required to create such a simple **for** loop, it must be used with caution. When macros are used with other operators, other macros, or other functions, unexpected program bugs can occur. For example, the above macro will not work at all with a pointer as the **var** argument, because **DO(*ptr,1,10)** would increment the pointer's value and not the value it points to (see section 2.7.1). This would probably result in a very strange number of cycles through the loop (if the loop ever terminated). As another example, consider the following **CUBE** macro, which will determine the cube of a variable:

```
#define CUBE(x)  (x)*(x)*(x)
```

This macro will work fine (although inefficiently) with **CUBE(i+j)**, since it would expand to **(i+j)*(i+j)*(i+j)**. However, **CUBE(i++)** expands to **(i++)*(i++)*(i++)**, resulting in **i** getting incremented three times instead of once. The resulting value would be $x(x+1)(x+2)$ not x^3.

The ternary conditional operator (see section 2.4.3) can be used with macro definitions to make fast implementations of the absolute value of a variable (**ABS**), the minimum of two variables (**MIN**), the maximum of two variables (**MAX**), and the integer rounded value of a floating-point variable (**ROUND**) as follows:

```
#define ABS(a)        (((a) < 0) ? (-a) : (a)
#define MAX(a,b)      (((a) > (b)) ? (a) : (b))
#define MIN(a,b)      (((a) < (b)) ? (a): (b))
#define ROUND(a)      (((a)<0)?(int)((a)-0.5):(int)((a)+0.5))
```

Note that each of the above macros is enclosed in parentheses so that it can be used freely in expressions without uncertainty about the order of operations. Parentheses are also required around each of the macro parameters, since these may contain operators as well as simple variables.

All of the macros defined so far have names that contain only capital letters. While this is not required, it does make it easy to separate macros from normal C keywords in programs where macros may be defined in one module and included (using the **#include** directive) in another. This practice of capitalizing all macro names and using lower case for variable and function names will be used in all programs in this book and on the accompanying disk.

2.7 POINTERS AND ARRAYS

A *pointer* is a variable that holds an address of some data, rather than the data itself. The use of pointers is usually closely related to manipulating (assigning or changing) the elements of an array of data. Pointers are used primarily for three purposes:

(1) To point to different data elements within an array

(2) To allow a program to create new variables while a program is executing (dynamic memory allocation)

(3) To access different locations in a data structure

The first two uses of pointers will be discussed in this section; pointers to data structures are considered in section 2.8.2.

2.7.1 Special Pointer Operators

Two *special pointer operators* are required to effectively manipulate pointers: the indirection operator (*****) and the address of operator (**&**). The indirection operator (*****) is used whenever the data stored at the address pointed to by a pointer is required, that is, whenever indirect addressing is required. Consider the following simple program:

```
main()
{
    int i,*ptr;
    i = 7;                    /* set the value of i */
    ptr = &i;                 /* point to address of i */
    printf("\n%d",i);         /* print i two ways */
    printf("\n%d",*ptr);
    *ptr = 11;                /* change i with pointer */
    printf("\n%d %d",*ptr,i);    /* print change */
}
```

This program declares that **i** is an integer variable and that **ptr** is a pointer to an integer variable. The program first sets **i** to 7 and then sets the pointer to the address of **i** by the statement **ptr=&i;**. The compiler assigns **i** and **ptr** storage locations somewhere in memory. At run time, **ptr** is set to the starting address of the integer variable **i**. The above program uses the function **printf** (see section 2.9.1) to print the integer value of i in two different ways—by printing the contents of the variable **i** (**printf("\n%d",i);**), and by using the indirection operator (**printf("\n%d",*ptr);**). The presence of the ***** operator in front of **ptr** directs the compiler to pass the value stored at the address

ptr to the **printf** function (in this case, 7). If only **ptr** were used, then the address assigned to **ptr** would be displayed instead of the value 7. The last two lines of the example illustrate indirect storage; the data at the **ptr** address is changed to 11. This results in changing the value of **i** only because **ptr** is pointing to the address of **i**.

An *array* is essentially a section of memory that is allocated by the compiler and assigned the name given in the declaration statement. In fact, the name given is nothing more than a fixed pointer to the beginning of the array. In C, the array name can be used as a pointer or it can be used to reference an element of the array (i.e., **a[2]**). If **a** is declared as some type of array then ***a** and **a[0]** are exactly equivalent. Furthermore, ***(a+i)** and **a[i]** are also the same (as long as **i** is declared as an integer), although the meaning of the second is often more clear. Arrays can be rapidly and sequentially accessed by using pointers and the increment operator (**++**). For example, the following three statements set the first 100 elements of the array **a** to 10:

```
int *pointer;
pointer = a;
for(i = 0; i < 100 ; i++) *pointer++ = 10;
```

On many computers this code will execute faster than the single statement **for(i=0; i<100; i++) a[i]=10;**, because the post increment of the pointer is faster than the array index calculation.

2.7.2 Pointers and Dynamic Memory Allocation

C has a set of four standard functions that allow the programmer to dynamically change the type and size of variables and arrays of variables stored in the computer's memory. C programs can use the same memory for different purposes and not waste large sections of memory on arrays only used in one small section of a program. In addition, **auto** variables are automatically allocated on the stack at the beginning of a function (or any section of code where the variable is declared within a pair of braces) and removed from the stack when a function is exited (or at the right brace, **}**). By proper use of **auto** variables (see section 2.5.2) and the dynamic memory allocation functions, the memory used by a particular C program can be very little more than the memory required by the program at every step of execution. This feature of C is especially attractive in multiuser environments where the product of the memory size required by a user and the time that memory is used ultimately determines the overall system performance. In many DSP applications, the proper use of dynamic memory allocation can enable a complicated DSP function to be performed with an inexpensive single chip signal processor with a small limited internal memory size instead of a more costly processor with a larger external memory.

Four standard functions are used to manipulate the memory available to a particular program (sometimes called the heap to differentiate it from the stack). **Malloc** allocates bytes of storage, **calloc** allocates items which may be any number of bytes long, **free** removes a previously allocated item from the heap, and **realloc** changes the size of a previously allocated item.

When using each function, the size of the item to be allocated must be passed to the function. The function then returns a pointer to a block of memory at least the size of the item or items requested. In order to make the use of the memory allocation functions portable from one machine to another, the built-in compiler macro **sizeof** must be used. For example:

```
int *ptr;
ptr = (int *) malloc(sizeof(int));
```

allocates storage for one integer and points the integer pointer, **ptr**, to the beginning of the memory block. On 32-bit machines this will be a four-byte memory block (or one word) and on 16-bit machines (such as the IBM PC) this will typically be only two bytes. Because **malloc** (as well as **calloc** and **realloc**) returns a character pointer, it must be cast to the integer type of pointer by the (**int ***) cast operator. Similarly, **calloc** and a pointer, **array**, can be used to define a 25-element integer array as follows:

```
int *array;
array = (int *) calloc(25,sizeof(int));
```

This statement will allocate an array of 25 elements, each of which is the size of an **int** on the target machine. The array can then be referenced by using another pointer (changing the pointer array is unwise, because it holds the position of the beginning of the allocated memory) or by an array reference such as **array[i]** (where **i** may be from 0 to 24). The memory block allocated by **calloc** is also initialized to zeros.

Malloc, **calloc**, and **free** provide a simple general purpose memory allocation package. The argument to **free** (cast as a character pointer) is a pointer to a block previously allocated by **malloc** or **calloc**; this space is made available for further allocation, but its contents are left undisturbed. Needless to say, grave disorder will result if the space assigned by **malloc** is overrun, or if some random number is handed to **free**. The function **free** has no return value, because memory is always assumed to be happily given up by the operating system.

Realloc changes the size of the block previously allocated to a new size in bytes and returns a pointer to the (possibly moved) block. The contents of the old memory block will be unchanged up to the lesser of the new and old sizes. **Realloc** is used less than **calloc** and **malloc**, because the size of an array is usually known ahead of time. However, if the size of the integer array of 25 elements allocated in the last example must be increased to 100 elements, the following statement can be used:

```
array = (int *) realloc( (char *)array, 100*sizeof(int));
```

Note that unlike **calloc**, which takes two arguments (one for the number of items and one for the item size), **realloc** works similar to **malloc** and takes the total size of the array in bytes. It is also important to recognize that the following two statements are not equivalent to the previous **realloc** statement:

```
free((char *)array);
array = (int *) calloc(100,sizeof(int));
```

These statements do change the size of the integer array from 25 to 100 elements, but do not preserve the contents of the first 25 elements. In fact, **calloc** will initialize all 100 integers to zero, while **realloc** will retain the first 25 and not set the remaining 75 array elements to any particular value.

Unlike **free**, which returns no value, **malloc**, **realloc**, and **calloc** return a null pointer (0) if there is no available memory or if the area has been corrupted by storing outside the bounds of the memory block. When **realloc** returns 0, the block pointed to by the original pointer may be destroyed.

2.7.3 Arrays of Pointers

Any of the C data types or pointers to each of the data types can be declared as an array. Arrays of pointers are especially useful in accessing large matrices. An array of pointers to 10 rows each of 20 integer elements can be dynamically allocated as follows:

```
int *mat[10];
int i;
for(i = 0 ; i < 10 ; i++) {
    mat[i] = (int *)calloc(20,sizeof(int));
    if(!mat[i]) {
        printf("\nError in matrix allocation\n");
        exit(1);
    }
}
```

In this code segment, the array of 10 integer pointers is declared and then each pointer is set to 10 different memory blocks allocated by 10 successive calls to **calloc**. After each call to **calloc**, the pointer must be checked to insure that the memory was available (**!mat[i]** will be true if **mat[i]** is null). Each element in the matrix **mat** can now be accessed by using pointers and the indirection operator. For example, ***(mat[i] + j)** gives the value of the matrix element at the **i**th row (0-9) and the **j**th column (0-19) and is exactly equivalent to **mat[i][j]**. In fact, the above code segment is equivalent (in the way **mat** may be referenced at least) to the array declaration **int mat[10][20];**, except that **mat[10][20]** is allocated as an **auto** variable on the stack and the above calls to **calloc** allocates the space for **mat** on the heap. Note, however, that when **mat** is allocated on the stack as an **auto** variable, it cannot be used with **free** or **realloc** and may be accessed by the resulting code in a completely different way.

The calculations required by the compiler to access a particular element in a two-dimensional matrix (by using **matrix[i][j]**, for example) usually take more instructions and more execution time than accessing the same matrix using pointers. This is especially true if many references to the same matrix row or column are required. However, depending on the compiler and the speed of pointer operations on the target machine, access to a two-dimensional array with pointers and simple pointers operands (even increment and decrement) may take almost the same time as a reference to a matrix such as

a[i][j]. For example, the product of two 100 x 100 matrices could be coded using two-dimensional array references as follows:

```
int a[100][100],b[100][100],c[100][100];     /* 3 matrices */
int i,j,k;                                    /* indices */

/* code to set up mat and vec goes here */

/* do matrix multiply c = a * b */
for(i = 0 ; i < 100 ; j++) {
    for(j = 0 ; j < 100 ; j++) {
        c[i][j] = 0;
        for(k = 0 ; k < 100 ; k++)
            c[i][j] += a[i][k] * b[k][j];
    }
}
```

The same matrix product could also be performed using arrays of pointers as follows:

```
int a[100][100],b[100][100],c[100][100];     /* 3 matrices */
int *aptr,*bptr,*cptr;                        /* pointers to a,b,c */
int i,j,k;                                    /* indicies */

/* code to set up mat and vec goes here */

/* do c = a * b */
for(i = 0 ; i < 100 ; i++) {
    cptr = c[i];
    bptr = b[0];
    for(j = 0 ; j < 100 ; j++) {
        aptr = a[i];
        *cptr = (*aptr++) * (*bptr++);
        for(k = 1 ; k < 100; k++) {
            *cptr += (*aptr++) * b[k][j];
        }
        cptr++;
    }
}
```

The latter form of the matrix multiply code using arrays of pointers runs 10 to 20 percent faster, depending on the degree of optimization done by the compiler and the capabilities of the target machine. Note that **c[i]** and **a[i]** are references to arrays of pointers each pointing to 100 integer values. Three factors help make the program with pointers faster:

(1) Pointer increments (such as ***aptr++**) are usually faster than pointer adds.

(2) No multiplies or shifts are required to access a particular element of each matrix.

(3) The first add in the inner most loop (the one involving **k**) was taken outside the loop (using pointers **aptr** and **bptr**) and the initialization of **c[i][j]** to zero was removed.

2.8 STRUCTURES

Pointers and arrays allow the same type of data to be arranged in a list and easily accessed by a program. Pointers also allow arrays to be passed to functions efficiently and dynamically created in memory. When unlike logically related data types must be manipulated, the use of several arrays becomes cumbersome. While it is always necessary to process the individual data types separately, it is often desirable to move all of the related data types as a single unit. The powerful C data construct called a *structure* allows new data types to be defined as a combination of any number of the standard C data types. Once the size and data types contained in a structure are defined (as described in the next section), the named structure may be used as any of the other data types in C. Arrays of structures, pointers to structures, and structures containing other structures may all be defined.

One drawback of the user-defined structure is that the standard operators in C do not work with the new data structure. Although the enhancements to C available with the C++ programming language do allow the user to define structure operators (see *The C++ Programming Language,* Stroustrup, 1986), the widely used standard C language does not support such concepts. Thus, functions or macros are usually created to manipulate the structures defined by the user. As an example, some of the functions and macros required to manipulate structures of complex floating-point data are discussed in section 2.8.3.

2.8.1 Declaring and Referencing Structures

A structure is defined by a structure template indicating the type and name to be used to reference each element listed between a pair of braces. The general form of an N-element structure is as follows:

```
struct tag_name {
    type1 element_name1;
    type2 element_name2;
              .
              .
              .
    typeN element_nameN;
        } variable_name;
```

In each case, **type1**, **type2**, . . . , **typeN** refer to a valid C data type (**char**, **int**, **float**, or **double** without any storage class descriptor) and **element_name1**, **element_name2**, . . . , **element_nameN** refer to the name of one of the elements of the data structure. The **tag_name** is an optional name used for referencing the structure later. The optional **variable_name**, or list of variable names, defines the names of the structures to be defined. The following structure template with a tag name of **record** defines a structure containing an integer called **length**, a **float** called **sample_rate**, a character pointer called **name**, and a pointer to an integer array called **data**:

```
struct record {
    int length;
    float sample_rate;
    char *name;
    int *data;
        };
```

This structure template can be used to declare a structure called voice as follows:

```
struct record voice;
```

The structure called **voice** of type **record** can then be initialized as follows:

```
voice.length = 1000;
voice.sample_rate = 10.e3;
voice.name = "voice signal";
```

The last element of the structure is a pointer to the data and must be set to the beginning of a 1000-element integer array (because length is 1000 in the above initialization). Each element of the structure is referenced with the form **struct_name.element**. Thus, the 1000-element array associated with the voice structure can be allocated as follows:

```
voice.data = (int *) calloc(1000,sizeof(int));
```

Similarly, the other three elements of the structure can be displayed with the following code segment:

```
printf("\nLength = %d",voice.length);
printf("\nSampling rate = %f",voice.sample_rate);
printf("\nRecord name = %s",voice.name);
```

A **typedef** statement can be used with a structure to make a user-defined data type and make declaring a structure even easier. The **typedef** defines an alternative name for the structure data type, but is more powerful than **#define**, since it is a compiler directive as opposed to a preprocessor directive. An alternative to the **record** structure is a **typedef** called **RECORD** as follows:

```
typedef struct record RECORD;
```

This statement essentially replaces all occurrences of **RECORD** in the program with the **struct record** definition thereby defining a new type of variable called **RECORD** that may be used to define the voice structure with the simple statement **RECORD voice;**.

The **typedef** statement and the structure definition can be combined so that the tag name **record** is avoided as follows:

```
typedef struct {
    int length;
    float sample_rate;
    char *name;
    int *data;
    } RECORD;
```

In fact, the **typedef** statement can be used to define a shorthand form of any type of data type including pointers, arrays, arrays of pointers, or another **typedef**. For example,

```
typedef char STRING[80];
```

allows 80-character arrays to be easily defined with the simple statement **STRING name1,name2;**. This shorthand form using the **typedef** is an exact replacement for the statement **char name1[80],name2[80];**.

2.8.2 Pointers to Structures

Pointers to structures can be used to dynamically allocate arrays of structures and efficiently access structures within functions. Using the **RECORD typedef** defined in the last section, the following code segment can be used to dynamically allocate a five-element array of **RECORD** structures:

```
RECORD *voices;
voices = (RECORD *) calloc(5,sizeof(RECORD));
```

These two statements are equivalent to the single-array definition **RECORD voices[5];** except that the memory block allocated by **calloc** can be deallocated by a call to the **free** function. With either definition, the length of each element of the array could be printed as follows:

```
int i;
for(i = 0 ; i < 5 ; i++)
    printf("\nLength of voice %d = %d",i,voices[i].length);
```

The **voices** array can also be accessed by using a pointer to the array of structures. If **voice_ptr** is a **RECORD** pointer (by declaring it with **RECORD *voice_ptr;**), then

(*voice_ptr).length could be used to give the length of the **RECORD** which was pointed to by **voice_ptr**. Because this form of pointer operation occurs with structures often in C, a special operator (**->**) was defined. Thus, **voice_ptr->length** is equivalent to **(*voice_ptr).length**. This shorthand is very useful when used with functions, since a local copy of a structure pointer is passed to the function. For example, the following function will print the length of each record in an array of **RECORD** of length **size:**

```
void print_record_length(RECORD *rec,int size)
{
    int i;
    for(i = 0 ; i < size ; i++) {
        printf("\nLength of record %d = %d",i,rec_>length);
       rec++;
    }
}
```

Thus, a statement like **print_record_length(voices,5);** will print the lengths stored in each of the five elements of the array of **RECORD** structures.

2.8.3 Complex Numbers

A complex number can be defined in C by using a structure of two **floats** and a **typedef** as follows:

```
typedef struct {
    float real;
    float imag;
        } COMPLEX;
```

Three complex numbers, **x**, **y**, and **z** can be defined using the above structure as follows:

```
COMPLEX x,y,z;
```

In order to perform the complex addition z = x + y without functions or macros, the following two C statements are required:

```
z.real = x.real + y.real;
z.imag = x.imag + y.imag;
```

These two statements are required because the C operator **+** can only work with the individual parts of the complex structure and not the structure itself. In fact, a statement involving any operator and a structure should give a compiler error. Assignment of any structure (like **z = x;**) works just fine, because only data movement is involved. A simple function to perform the complex addition can be defined as follows:

```
COMPLEX cadd(COMPLEX a,COMPLEX b)  /* pass by value */
{
    COMPLEX sum;     /* define return value */
    sum.real = a.real + b.real;
    sum.imag = a.imag + b.imag;
    return(sum);
}
```

This function passes the value of the **a** and **b** structures, forms the sum of **a** and **b**, and then returns the complex summation (some compilers may not allow this method of passing structures by value, thus requiring pointers to each of the structures). The **cadd** function may be used to set **z** equal to the sum of **x** and **y** as follows:

```
z = cadd(x,y);
```

The same complex sum can also be performed with a rather complicated single line macro defined as follows:

```
#define CADD(a,b)\
  (C_t.real=a.real+b.real,C_t.imag=a.imag+b.imag,C_t)
```

This macro can be used to replace the **cadd** function used above as follows:

```
COMPLEX C_t;
z = CADD(z,y);
```

This **CADD** macro works as desired because the macro expands to three operations separated by commas. The one-line macro in this case is equivalent to the following three statements:

```
C_t.real = x.real + y.real;
C_t.imag = x.imag + y.real;
z = C_t;
```

The first two operations in the macro are the two sums for the real and imaginary parts. The sums are followed by the variable **C_t** (which must be defined as **COMPLEX** before using the macro). The expression formed is evaluated from left to right and the whole expression in parentheses takes on the value of the last expression, the complex structure **C_t**, which gets assigned to **z** as the last statement above shows.

The complex add macro **CADD** will execute faster than the **cadd** function because the time required to pass the complex structures **x** and **y** to the function, and then pass the sum back to the calling program, is a significant part of the time required for the function call. Unfortunately, the complex add macro cannot be used in the same manner as the function. For example:

```
COMPLEX a,b,c,d;
d = cadd(a,cadd(b,c));
```

will form the sum **d=a+b+c;** as expected. However, the same format using the **CADD** macro would cause a compiler error, because the macro expansion performed by the C preprocessor results in an illegal expression. Thus, the **CADD** may only be used with simple single-variable arguments. If speed is more important than ease of programming, then the macro form should be used by breaking complicated expressions into simpler two-operand expressions. Numerical C extensions to the C language support complex numbers in an optimum way and are discussed in section 2.10.1.

2.9 COMMON C PROGRAMMING PITFALLS

The following sections describe some of the more common errors made by programmers when they first start coding in C and give a few suggestions how to avoid them.

2.9.1 Array Indexing

In C, all array indices start with zero rather than one. This makes the last index of a N long array N-1. This is very useful in digital signal processing, because many of the expressions for filters, z-transforms, and FFTs are easier to understand and use with the index starting at zero instead of one. For example, the FFT output for $k = 0$ gives the zero frequency (DC) spectral component of a discrete time signal. A typical indexing problem is illustrated in the following code segment, which is intended to determine the first 10 powers of 2 and store the results in an array called **power2:**

```
int power2[10];
int i,p;
p = 1;
for (i = 1 ; i<= 10 ; i++) {
    power2[i] = p;
    p = 2*p;
}
```

This code segment will compile well and may even run without any difficulty. The problem is that the **for** loop index **i** stops on **i=10**, and **power2[10]** is not a valid index to the **power2** array. Also, the **for** loop starts with the index 1 causing **power2[0]** to not be initialized. This results in the first power of two (2^0, which should be stored in **power2[0]**) to be placed in **power2[1]**. One way to correct this code is to change the for loop to read **for(i = 0; i<10; i++)**, so that the index to **power2** starts at 0 and stops at 9.

2.9.2 Failure to Pass-by-Address

This problem is most often encountered when first using **scanf** to read in a set of variables. If **i** is an integer (declared as **int i;**), then a statement like **scanf("%d",i);** is wrong because **scanf** expects the address of (or pointer to) the location to store the integer that is read by **scanf**. The correct statement to read in the integer **i** is **scanf("%d",&i);**, where the address of operator (**&**) was used to point to the address of **i** and pass the address to **scanf** as required. On many compilers these types of errors can be detected and avoided by using function prototypes (see section 2.5.3) for all user written functions and the appropriate include files for all C library functions. By using function prototypes, the compiler is informed what type of variable the function expects and will issue a warning if the specified type is not used in the calling program. On many UNIX systems, a C program checker called LINT can be used to perform parameter-type checking, as well as other syntax checking.

2.9.3 Misusing Pointers

Because pointers are new to many programmers, the misuse of pointers in C can be particularly difficult, because most C compilers will not indicate any pointer errors. Some compilers issue a warning for some pointer errors. Some pointer errors will result in the programs not working correctly or, worse yet, the program may seem to work, but will not work with a certain type of data or when the program is in a certain mode of operation. On many small single-user systems (such as the IBM PC), misused pointers can easily result in writing to memory used by the operating system, often resulting in a system crash and requiring a subsequent reboot.

There are two types of pointer abuses: setting a pointer to the wrong value (or not initializing it at all) and confusing arrays with pointers. The following code segment shows both of these problems:

```
char *string;
char msg[10];
int i;
printf("\nEnter title");
scanf("%s",string);
i = 0;
while(*string != ' ') {
    i++;
    string++;
}
msg="Title = ";
printf("%s %s %d before space", msg, string,i);
```

The first three statements declare that memory be allocated to a pointer variable called **string**, a 10-element **char** array called **msg** and an integer called **i**. Next, the user is asked to enter a title into the variable called **string**. The **while** loop is intended to search for the first space in the string and the last **printf** statement is intended to display the string and the number of characters before the first space.

There are three pointer problems in this program, although the program will compile with only one fatal error (and a possible warning). The fatal error message will reference the **msg="Title =";** statement. This line tells the compiler to set the address of the **msg** array to the constant string **"Title =".** This is not allowed so the error

"Lvalue required" (or something less useful) will be produced. The role of an array and a pointer have been confused and the **msg** variable should have been declared as a pointer and used to point to the constant string **"Title ="**, which was already allocated storage by the compiler.

The next problem with the code segment is that **scanf** will read the string into the address specified by the argument **string**. Unfortunately, the value of **string** at execution time could be anything (some compilers will set it to zero), which will probably not point to a place where the title string could be stored. Some compilers will issue a warning indicating that the pointer called **string** may have been used before it was defined. The problem can be solved by initializing the string pointer to a memory area allocated for storing the title string. The memory can be dynamically allocated by a simple call to **calloc** as shown in the following corrected code segment:

```
char *string,*msg;
int i;
string=calloc(80,sizeof(char));
printf("\nEnter title");
scanf("%s", string);
i = 0;
while(*string != ' ') {
     i++;
     string++;
}
msg="Title =";
printf("%s %s %d before space",msg,string,i);
```

The code will now compile and run but will not give the correct response when a title string is entered. In fact, the first characters of the title string before the first space will not be printed because the pointer **string** was moved to this point by the execution of the **while** loop. This may be useful for finding the first space in the **while** loop, but results in the address of the beginning of the string being lost. It is best not to change a pointer which points to a dynamically allocated section of memory. This pointer problem can be fixed by using another pointer (called **cp**) for the **while** loop as follows:

```
char *string,*cp,*msg;
int i;
string=calloc(80,sizeof(char));
printf("\nEnter title");
scanf("%s",string);
i = 0;
cp = string;
while(*cp != ' ') {
     i++;
     cp++;
}
msg="Title =";
printf("%s %s %d before space", msg,string,i);
```

Another problem with this program segment is that if the string entered contains no spaces, then the **while** loop will continue to search through memory until it finds a space. On a PC, the program will almost always find a space (in the operating system perhaps) and will set **i** to some large value. On larger multiuser systems, this may result in a fatal run-time error because the operating system must protect memory not allocated to the program. Although this programming problem is not unique to C, it does illustrate an important characteristic of pointers—pointers can and will point to any memory location without regard to what may be stored there.

2.10 NUMERICAL C EXTENSIONS

Some ANSI C compilers designed for DSP processors are now available with numeric C extensions. These language extensions were developed by the ANSI NCEG (Numeric C Extensions Group), a working committee reporting to ANSI X3J11. This section gives an overview of the *Numerical C* language recommended by the ANSI standards committee. Numerical C has several features of interest to DSP programmers:

(1) Fewer lines of code are required to perform vector and matrix operations.

(2) Data types and operators for complex numbers (with real and imaginary components) are defined and can be optimized by the compiler for the target processor. This avoids the use of structures and macros as discussed in section 2.8.3.

(3) The compiler can perform better optimizations of programs containing iteration which allows the target processor to complete DSP tasks in fewer instruction cycles.

2.10.1 Complex Data Types

Complex numbers are supported using the keywords **complex**, **creal**, **cimag**, and **conj**. These keywords are defined when the header file **complex.h** is included. There are six integer complex types and three floating-point complex types, defined as shown in the following example:

```
short int complex i;
int complex j;
long int complex k;
unsigned short int complex ui;
unsigned int complex uj;
unsigned long int complex uk;
float complex x;
double complex y;
long double complex z;
```

The real and imaginary parts of the complex types each have the same representations as the type defined without the complex keyword. Complex constants are represented as a sum of a real constant and an imaginary constant, which is defined by using the suffix **i** after the imaginary part of the number. For example, initialization of complex numbers is performed as follows:

```
short int complex i = 3 + 2i;
float complex x[3] = {1.0+2.0i, 3.0i, 4.0};
```

The following operators are defined for complex types: **&** (address of), ***** (point to complex number), **+** (add), – (subtract), ***** (multiply), **/** (divide). Bitwise, relational, and logical operators are not defined. If any one of the operands are **complex**, the other operands will be converted to **complex**, and the result of the expression will be **complex**. The **creal** and **cimag** operators can be used in expressions to access the real or imaginary part of a complex variable or constant. The **conj** operator returns the complex conjugate of its complex argument. The following code segment illustrates these operators:

```
float complex a,b,c;
creal(a)=1.0;
cimag(a)=2.0;
creal(b)=2.0*cimag(a);
cimag(b)=3.0;
c=conj(b);          /* c will be 4 - 3i */
```

2.10.2 Iteration Operators

Numerical C offers *iterators* to make writing mathematical expressions that are computed iteratively more simple and more efficient. The two new keywords are **iter** and **sum**. Iterators are variables that expand the execution of an expression to iterate the expression so that the iteration variable is automatically incremented from 0 to the value of the iterator. This effectively places the expression inside an efficient **for** loop. For example, the following three lines can be used to set the 10 elements of array **ix** to the integers 0 to 9:

```
iter I=10;
int ix[10];
ix[I]=I;
```

The **sum** operator can be used to represent the sum of values computed from values of an iterator. The argument to **sum** must be an expression that has a value for each of the iterated variables, and the order of the iteration cannot change the result. The following code segment illustrates the **sum** operator:

```
float a[10],b[10],c[10],d[10][10],e[10][10],f[10][10];
float s;
iter I=10, J=10, K=10;
s=sum(a[I]);      /* computes the sum of a into s */
```

```
b[J]=sum(a[I]);   /* sum of a calculated 10 times and stored
                     in the elements of b */
c[J]=sum(d[I][J]);  /* computes the sum of the column
                       elements of d, the statement is
                       iterated over J */
s=sum(d[I][J]);     /* sums all the elements in d */
f[I][J]=sum(d[I][K]*e[K][J]); /* matrix multiply */
c[I]=sum(d[I][K]*a[K]);       /* matrix * vector */
```

2.11 COMMENTS ON PROGRAMMING STYLE

The four common measures of good DSP software are *reliability, maintainability, extensibility,* and *efficiency.*

A reliable program is one that seldom (if ever) fails. This is especially important in DSP because tremendous amounts of data are often processed using the same program. If the program fails due to one sequence of data passing through the program, it may be difficult, or impossible, to ever determine what caused the problem.

Since most programs of any size will occasionally fail, a maintainable program is one that is easy to fix. A truly maintainable program is one that can be fixed by someone other than the original programmer. It is also sometimes important to be able to maintain a program on more than one type of processor, which means that in order for a program to be truly maintainable, it must be portable.

An extensible program is one that can be easily modified when the requirements change, new functions need to be added, or new hardware features need to be exploited.

An efficient program is often the key to a successful DSP implementation of a desired function. An efficient DSP program will use the processing capabilities of the target computer (whether general purpose or dedicated) to minimize the execution time. In a typical DSP system this often means minimizing the number of operations per input sample or maximizing the number of operations that can be performed in parallel. In either case, minimizing the number of operations per second usually means a lower overall system cost as fast computers typically cost more than slow computers. For example, it could be said that the FFT algorithm reduced the cost of speech processing (both implementation cost and development cost) such that inexpensive speech recognition and generation processors are now available for use by the general public.

Unfortunately, DSP programs often forsake maintainability and extensibility for efficiency. Such is the case for most currently available programmable signal processing integrated circuits. These devices are usually programmed in assembly language in such a way that it is often impossible for changes to be made by anyone but the original programmer, and after a few months even the original programmer may have to rewrite the program to add additional functions. Often a compiler is not available for the processor or the processor's architecture is not well suited to efficient generation of code from a compiled language. The current trend in programmable signal processors appears to be toward high-level languages. In fact, many of the DSP chip manufacturers are supplying C compilers for their more advanced products.

2.11.1 Software Quality

The four measures of software quality (reliability, maintainability, extensibility, and efficiency) are rather difficult to quantify. One almost has to try to modify a program to find out if it is maintainable or extensible. A program is usually tested in a finite number of ways much smaller than the millions of input data conditions. This means that a program can be considered reliable only after years of bug-free use in many different environments.

Programs do not acquire these qualities by accident. It is unlikely that good programs will be intuitively created just because the programmer is clever, experienced, or uses lots of comments. Even the use of structured-programming techniques (described briefly in the next section) will not assure that a program is easier to maintain or extend. It is the author's experience that the following five coding situations will often lessen the software quality of DSP programs:

(1) Functions that are too big or have several purposes
(2) A main program that does not use functions
(3) Functions that are tightly bound to the main program
(4) Programming "tricks" that are always poorly documented
(5) Lack of meaningful variable names and comments

An *oversized function* (item 1) might be defined as one that exceeds two pages of source listing. A function with more than one purpose lacks strength. A function with one clearly defined purpose can be used by other programs and other programmers. Functions with many purposes will find limited utility and limited acceptance by others. All of the functions described in this book and contained on the included disk were designed with this important consideration in mind. Functions that have only one purpose should rarely exceed one page. This is not to say that all functions will be smaller than this. In time-critical DSP applications, the use of in-line code can easily make a function quite long but can sometimes save precious execution time. It is generally true, however, that big programs are more difficult to understand and maintain than small ones.

A *main program* that does not use functions (item 2) will often result in an extremely long and hard-to-understand program. Also, because complicated operations often can be independently tested when placed in short functions, the program may be easier to debug. However, taking this rule to the extreme can result in functions that are *tightly bound* to the main program, violating item 3. A function that is tightly bound to the rest of the program (by using too many global variables, for example) weakens the entire program. If there are lots of tightly coupled functions in a program, maintenance becomes impossible. A change in one function can cause an undesired, unexpected change in the rest of the functions.

Clever programming tricks (item 4) should be avoided at all costs as they will often not be reliable and will almost always be difficult for someone else to understand (even with lots of comments). Usually, if the program timing is so close that a trick must be used, then the wrong processor was chosen for the application. Even if the programming trick solves a particular timing problem, as soon as the system requirements change (as they almost always do), a new timing problem without a solution may soon develop.

A program that does not use *meaningful variables and comments* (item 5) is guaranteed to be very difficult to maintain. Consider the following valid C program:

```
main(){int _o_oo_,_ooo;for(_o_oo__=2;;__o__o__++)
{for(__ooo_=2;_o__oo__%__ooo_!=0;__ooo_++;
if(__ooo_==_o_oo__)printf("\n%d",_o_oo__);}}
```

Even the most experienced C programmer would have difficulty determining what this three-line program does. Even after running such a poorly documented program, it may be hard to determine how the results were obtained. The following program does exactly the same operations as the above three lines but is easy to follow and modify:

```
main()
{
  int prime_test,divisor;
/* The outer for loop trys all numbers >1 and the inner
for loop checks the number tested for any divisors
less than itself. */
  for(prime_test = 2 ; ; prime_test++) {
    for(divisor = 2 ; prime_test % divisor != 0 ; divisor++);
    if(divisor == prime_test) printf("\n%d",prime_test);
  }
}
```

It is easy for anyone to discover that the above well-documented program prints a list of prime numbers, because the following three documentation rules were followed:

(1) Variable names that are meaningful in the context of the program were used. Avoid variable names such as **x**,**y**,**z** or **i**,**j**,**k**, unless they are simple indexes used in a very obvious way, such as initializing an entire array to a constant.

(2) Comments preceded each major section of the program (the above program only has one section). Although the meaning of this short program is fairly clear without the comments, it rarely hurts to have too many comments. Adding a blank line between different parts of a program also sometimes improves the readability of a program because the different sections of code appear separated from each other.

(3) Statements at different levels of nesting were indented to show which control structure controls the execution of the statements at a particular level. The author prefers to place the right brace (**{**) with the control structure (**for**, **while**, **if**, etc.) and to place the left brace (**}**) on a separate line starting in the same column as the beginning of the corresponding control structure. The exception to this practice is in function declarations where the right brace is placed on a separate line after the argument declarations.

2.11.2 Structured Programming

Structured programming has developed from the notion that any algorithm, no matter how complex, can be expressed by using the programming-control structures *if-else, while,* and *sequence.* All programming languages must contain some representation of these three fundamental control structures. The development of structured programming revealed that if a program uses these three control structures, then the logic of the program can be read and understood by beginning at the first statement and continuing downward to the last. Also, all programs could be written without goto statements. Generally, structured-programming practices lead to code that is easier to read, easier to maintain, and even easier to write.

The C language has the three basic control structures as well as three additional structured-programming constructs called *do-while, for,* and *case.* The additional three control structures have been added to C and most other modern languages because they are convenient, they retain the original goals of structured programming, and their use often makes a program easier to comprehend.

The *sequence* control structure is used for operations that will be executed once in a function or program in a fixed sequence. This structure is often used where speed is most important and is often referred to as in-line code when the sequence of operations are identical and could be coded using one of the other structures. Extensive use of in-line code can obscure the purpose of the code segment.

The *if-else* control structure in C is the most common way of providing conditional execution of a sequence of operations based on the result of a logical operation. Indenting of different levels of **if** and **else** statements (as shown in the example in section 2.4.1) is not required; it is an expression of C programming style that helps the readability of the if-else control structure. Nested **while** and **for** loops should also be indented for improved readability (as illustrated in section 2.7.3).

The *case* control structure is a convenient way to execute one of a series of operations based upon the value of an expression (see the example in section 2.4.2). It is often used instead of a series of if-else structures when a large number of conditions are tested based upon a common expression. In C, the **switch** statement gives the expression to test and a series of **case** statements give the conditions to match the expression. A **default** statement can be optionally added to execute a sequence of operations if none of the listed conditions are met.

The last three control structures (while, do-while, and for) all provide for repeating a sequence of operations a fixed or variable number of times. These loop statements can make a program easy to read and maintain. The **while** loop provides for the iterative execution of a series of statements as long as a tested condition is true; when the condition is false, execution continues to the next statement in the program. The **do-while** control structure is similar to the **while** loop, except that the sequence of statements is executed at least once. The **for** control structure provides for the iteration of statements with automatic modification of a variable and a condition that terminates the iterations. **For** loops are more powerful in C than most languages. C allows for any initializing statement, any iterating statement and any terminating statement. The three statements do not need to be related and any of them can be a null statement or multiple statements. The following three examples of **while**, **do-while**, and **for** loops all calculate the power of two of an integer **i** (assumed to be greater than 0) and set the result to **k**. The **while** loop is as follows:

```
k = 2;   /* while loop k=2**i */
while(i > 0) {
    k = 2*k;
    i--;
}
```

The **do-while** loop is as follows:

```
k = 1;   /* do-while loop k = 2**i */
do {
    k = 2*k;
    i--;
} while(i > 0);
```

The **for** loop is as follows:

```
for(k = 2 ; i > 1 ; i--)
    k = 2*k; /* for loop k=2**i */
```

Which form of loop to use is a personal matter. Of the three equivalent code segments shown above, the **for** loop and the **while** loop both seem easy to understand and would probably be preferred over the **do-while** construction.

The C language also offers several extensions to the six structured programming control structures. Among these are **break**, **continue**, and **goto** (see section 2.4.5). **Break** and **continue** statements allow the orderly interruption of events that are executing inside of loops. They can often make a complicated loop very difficult to follow, because more than one condition may cause the iterations to stop. The infamous **goto** statement is also included in C. Nearly every language designer includes a **goto** statement with the advice that it should never be used along with an example of where it might be useful.

The program examples in the following chapters and the programs contained on the enclosed disk were developed by using structured-programming practices. The code can be read from top to bottom, there are no **goto** statements, and the six accepted control structures are used. One requirement of structured programming that was not adopted throughout the software in this book is that each program and function have only one entry and exit point. Although every function and program does have only one entry point (as is required in C), many of the programs and functions have multiple exit points. Typically, this is done in order to improve the readability of the program. For example, error conditions in a main program often require terminating the program prematurely after displaying an error message. Such an error-related exit is performed by calling the C library function **exit(n)** with a suitable error code, if desired. Similarly, many of the functions have more than one return statement as this can make the logic in a function much easier to program and in some cases more efficient.

2.12 REFERENCES

FEUER, A.R. (1982). *The C Puzzle Book.* Englewood Cliffs, NJ: Prentice Hall.

KERNIGHAM, B. and PLAUGER, P. (1978). *The Elements of Programming Style.* New York: McGraw-Hill.

KERNIGHAN, B.W. and RITCHIE, D.M. (1988). *The C Programming Language* (2nd ed.). Englewood Cliffs, NJ: Prentice Hall.

PRATA, S. (1986). *Advanced C Primer++.* Indianapolis, IN: Howard W. Sams and Co.

PURDUM, J. and LESLIE, T.C. (1987). *C Standard Library.* Indianapolis, IN: Que Co.

ROCHKIND, M.J. (1988). *Advanced C Programming for Displays.* Englewood Cliffs, NJ: Prentice Hall.

STEVENS, A. (1986). *C Development Tools for the IBM PC.* Englewood Cliffs, NJ: Prentice Hall.

STROUSTRUP, B. (1986). *The C++ Programming Language.* Reading, MA: Addison-Wesley.

WAITE, M., PRATA, S. and MARTIN, D. (1987). *C Primer Plus.* Indianapolis, IN: Howard W. Sams and Co.

CHAPTER 3

DSP MICROPROCESSORS IN EMBEDDED SYSTEMS

The term embedded system is often used to refer to a processor and associated circuits required to perform a particular function that is not the sole purpose of the overall system. For example, a keyboard controller on a computer system may be an embedded system if it has a processor that handles the keyboard activity for the computer system. In a similar fashion, digital signal processors are often embedded in larger systems to perform specialized DSP operations to allow the overall system to handle general purpose tasks. A special purpose processor used for voice processing, including analog-to-digital (A/D) and digital-to-analog (D/A) converters, is an embedded DSP system when it is part of a personal computer system. Often this type of DSP runs only one application (perhaps speech synthesis or recognition) and is not programmed by the end user. The fact that the processor is embedded in the computer system may be unknown to the end user.

A DSP's data format, either fixed-point or floating-point, determines its ability to handle signals of differing precision, dynamic range, and signal-to-noise ratios. Also, ease-of-use and software development time are often equally important when deciding between fixed-point and floating-point processors. Floating-point processors are often more expensive than similar fixed-point processors but can execute more instructions per second. Each instruction in a floating-point processor may also be more complicated, leading to fewer cycles per DSP function. DSP microprocessors can be classified as fixed-point processors if they can only perform fixed-point multiplies and adds, or as floating-point processors if they can perform floating-point operations.

The precision of a particular class of A/D and D/A converters (classified in terms of cost or maximum sampling rate) has been slowly increasing at a rate of about one bit every two years. At the same time the speed (or maximum sampling rate) has also been increasing. The dynamic range of many algorithms is higher at the output than at the input and intermediate results are often not constrained to any particular dynamic range. This requires that intermediate results be scaled using a shift operator when a fixed-point DSP is used. This will require more cycles for a particular algorithm in fixed-point than on an equal floating-point processor. Thus, as the A/D and D/A requirements for a particular application require higher speeds and more bits, a fixed-point DSP may need to be replaced with a faster processor with more bits. Also, the fixed-point program may require extensive modification to accommodate the greater precision.

In general, floating-point DSPs are easier to use and allow a quicker time-to-market than processors that do not support floating-point formats. The extent to which this is true depends on the architecture of the floating-point processor. High-level language programmability, large address spaces, and wide dynamic range associated with floating-point processors allow system development time to be spent on algorithms and signal processing problems rather than assembly coding, code partitioning, quantization error, and scaling. In the remainder of this chapter, floating-point digital signal processors and the software required to develop DSP algorithms are considered in more detail.

3.1 TYPICAL FLOATING-POINT DIGITAL SIGNAL PROCESSORS

This section describes the general properties of the following three floating-point DSP processor families: AT&T DSP32C and DSP3210, Analog Devices ADSP-21020 and ADSP-21060, and Texas Instruments TMS320C30 and TMS320C40. The information was obtained from the manufacturers' data books and manuals and is believed to be an accurate summary of the features of each processor and the development tools available. Detailed information should be obtained directly from manufacturers, as new features are constantly being added to their DSP products. The features of the three processors are summarized in sections 3.1.1, 3.1.2, and 3.1.3.

The execution speed of a DSP algorithm is also important when selecting a processor. Various basic building block DSP algorithms are carefully optimized in assembly language by the processor's manufacturer. The time to complete a particular algorithm is often called a benchmark. Benchmark code is always in assembly language (sometimes without the ability to be called by a C function) and can be used to give a general measure of the maximum signal processing performance that can be obtained for a particular processor. Typical benchmarks for the three floating-point processor families are shown in the following table. Times are in microseconds based the on highest speed processor available at publication time.

		DSP32C DSP3210	ADSP21020 ADSP21060	TMS320C30	TMS320C40
Maximum Instruction Cycle Speed (MIPs)		20	40	20	30
1024 Complex	cycles	161311*	19245	40457	38629
FFT with bitreverse	time	2016.4	481.13	2022.85	1287.63
FIR Filter	cycles	187*	44	45	42
(35 Taps)	time	2.34	1.1	2.25	1.4
IIR Filter	cycles	85*	14	23	21
(2 Biquads)	time	1.06	0.35	1.15	0.7
4×4 * 4×1	cycles	80*	24	58	37
Matrix Multiply	time	1.0	0.6	2.9	1.23

*Cycle counts for DSP32C and DSP3210 are clock cycles including wait states (1 instruction = 4 clock cycles).

3.1.1 AT&T DSP32C and DSP3210

Figure 3.1 shows a block diagram of the DSP32C microprocessor manufactured by AT&T (Allentown, PA). The following is a brief description of this processor provided by AT&T.

The DSP32C's two execution units, the control arithmetic unit (CAU) and the data arithmetic unit (DAU), are used to achieve the high throughput of 20 million instructions per second (at the maximum clock speed of 80 MHz). The CAU performs 16- or 24-bit integer arithmetic and logical operations, and provides data move and control capabilities. This unit includes 22 general purpose registers. The DAU performs 32-bit floating-point arithmetic for signal processing functions. It includes a 32-bit floating-point multiplier, a 40-bit floating-point adder, and four 40-bit accumulators. The multiplier and the adder work in parallel to perform 25 million floating-point computations per second. The DAU also incorporates special-purpose hardware for data-type conversions.

On-chip memory includes 1536 words of RAM. Up to 16 Mbytes of external memory can be directly addressed by the external memory interface that supports wait states and bus arbitration. All memory can be addressed as 8-, 16- or 32-bit data, with the 32-bit data being accessed at the same speed as 8- or 16-bit data.

The DSP32C has three I/O ports: an external memory port, a serial port and a 16-bit parallel port. In addition to providing access to commercially available memory, the external memory interface can be used for memory mapped I/O. The serial port can interface to a time division multiplexed (TDM) stream, a codec, or another DSP32C. The parallel port provides an interface to an external microprocessor. In summary, some of the key features of the DSP32C follow.

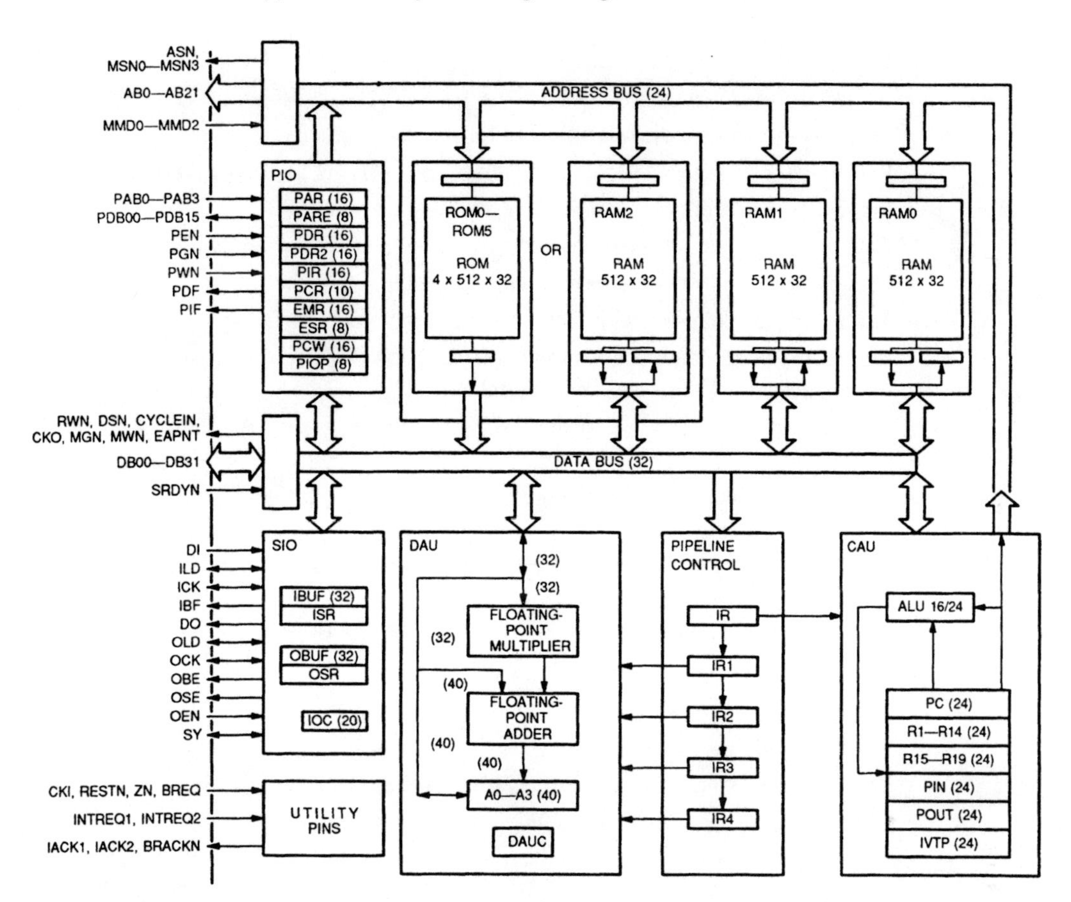

LEGEND*:

A0—A3	Accumulators 0—3	**ISR**	Input shift register	**PDR2**	PIO data register 2
ALU	Arithmetic logic unit	**IVTP**	Interrupt vector table pointer	**PIN**	Serial DMA input pointer
CAU	Control arithmetic unit	**OBUF**	Output buffer	**PIO**	Parallel I/O unit
DAU	Data arithmetic unit	**OSR**	Output shift register	**PIOP**	Parallel I/O port register
DAUC	DAU control register	**PAR**	PIO address register	**PIR**	PIO interrupt register
EMR	Error mask register	**PARE**	PIO address register extended	**POUT**	Serial DMA output pointer
ESR	Error source register	**PC**	Program counter	**R1—R19**	Registers 1—19
IBUF	Input buffer	**PCR**	PIO control register	**RAM**	Read/write memory
IOC	Input/output control register	**PCW**	Processor control word	**ROM**	Read-only memory
IR	Instruction register	**PDR**	PIO data register	**SIO**	Serial I/O unit
IR1—IR4	Instruction register pipeline				

* For a detailed description, see Architecture.

FIGURE 3.1 Block diagram of DSP32C processor (Courtesy AT&T).

KEY FEATURES

- 63 instructions, 3-stage pipeline
- Full 32-bit floating-point architecture for increased precision and dynamic range
- Faster application evaluation and development yielding increased market leverage
- Single cycle instructions, eliminating multicycle branches
- Four memory accesses/instruction for exceptional memory bandwidth
- Easy-to-learn and read C-like assembly language
- Serial and parallel ports with DMA for clean external interfacing
- Single-instruction data conversion for IEEE P754 floating-point, 8-, 16-, and 24-bit integers, μ-law, A-Law and linear numbers
- Fully vectored interrupts with hardware context saving up to 2 million interrupts per second
- Byte-addressable memory efficiently storing 8- and 16-bit data
- Wait-states of 1/4 instruction increments and two independent external memory speed partitions for flexible, cost-efficient memory configurations

DESCRIPTION OF AT&T HARDWARE DEVELOPMENT HARDWARE

- Full-speed operation of DSP32C (80 MHz)
- Serial I/O through an AT&T T7520 High Precision Codec with mini-BNC connectors for analog input and output
- Provision for external serial I/O through a 34-pin connector
- Upper 8-bits of DSP32C parallel I/O port access via connector
- DSP32C Simulator providing an interactive interface to the development system.
- Simulator-controlled software breakpoints; examining and changing memory contents or device registers

Figure 3.2 shows a block diagram of the DSP3210 microprocessor manufactured by AT&T. The following is a brief description of this processor provided by AT&T.

AT&T DSP3210 DIGITAL SIGNAL PROCESSOR FEATURES

- 16.6 million instructions per second (66 MHz clock)
- 63 instructions, 3-stage pipeline
- Two 4-kbyte RAM blocks, 1-kbyte boot ROM
- 32-bit barrel shifter and 32-bit timer
- Serial I/O port, 2 DMA channels, 2 external interrupts
- Full 32-bit floating-point arithmetic for fast, efficient software development
- C-like assembly language for ease of programming
- All single-cycle instructions; four memory accesses/instruction cycle
- Hardware context save and single-cycle PC relative addressing

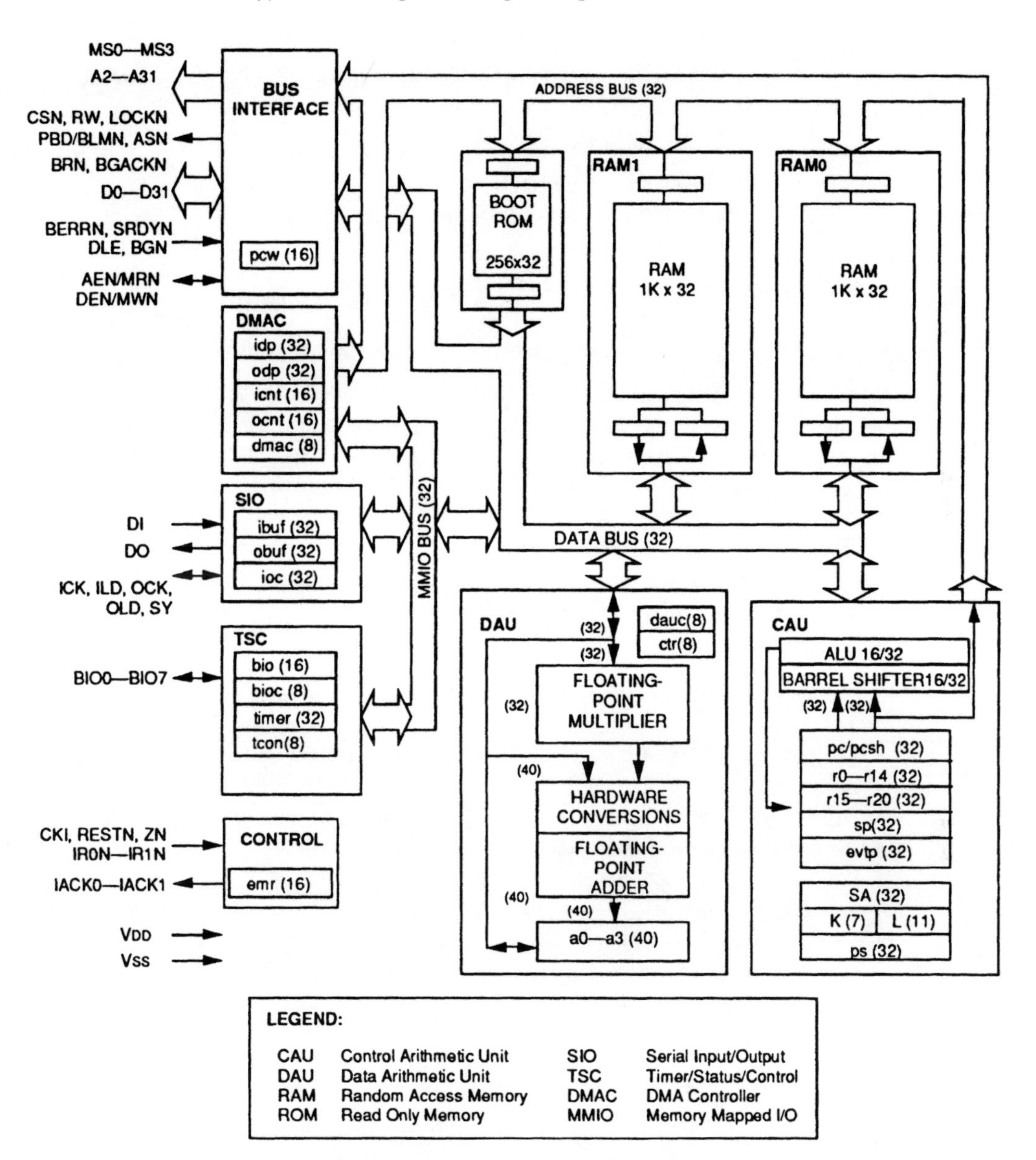

FIGURE 3.2 Block diagram of DSP3210 processor (Courtesy AT&T).

- Microprocessor bus compatibility (The DSP3210 is designed for efficient bus master designs. This allows the DSP3210 to be easily incorporated into microprocessor-based designs
- 32-bit, byte-addressable address space allowing the DSP3210 and a microprocessor to share the same address space and to share pointer values.
- Retry, relinquish/retry, and bus error support
- Page mode DRAM support
- Direct support for both Motorola and Intel signaling

AT&T DSP3210 FAMILY HARDWARE DEVELOPMENT SYSTEM DESCRIPTION

The MP3210 implements one or two AT&T DSP3210 32-bit floating-point DSPs with a comprehensive mix of memory, digital I/O and professional audio signal I/O. The MP3210 holds up to 2 Mbytes of dynamic RAM (DRAM). The DT-Connect interface enables real-time video I/O. MP3210 systems include: the processor card; C Host drivers with source code; demos, examples and utilities, with source code; User's Manual; and the AT&T DSP3210 Information Manual. DSP3210 is the low-cost Multimedia Processor of Choice. New features added to the DSP3210 are briefly outlined below.

DSP3210 FEATURES	USER BENEFITS
• Speeds up to 33 MFLOPS	The high performance and large on-chip memory space enable
• 2k × 32 on-chip RAM	fast, efficient processing of complex algorithms.
• Full, 32-bit, floating-point	Ease of programming/higher performance.
• All instructions are single cycle	
• Four memory accesses per instruction cycle	Higher performance.
• Microprocessor bus compatibility	Designed for efficient bus master.
• 32-bit byte-addressable designs.	This allows the DSP3210 address space to
• Retry, relinquish/retry	easily be incorporated into μP
• error support	bus-based designs. The 32-bit, byte-
• Boot ROM	addressable space allows the
• Page mode DRAM support	DSP3210 and a μP to share the same
• Directly supports 680X0 and 80X86 signaling	address space and to share pointer values as well.

3.1.2 Analog Devices ADSP-210XX

Figure 3.3 shows a block diagram of the ADSP-21020 microprocessor manufactured by Analog Devices (Norwood, MA). The ADSP-21060 core processor is similar to the ADSP-21020. The ADSP-21060 (Figure 3.4) also includes a large on-chip memory, a DMA controller, serial ports, link ports, a host interface and multiprocessing features

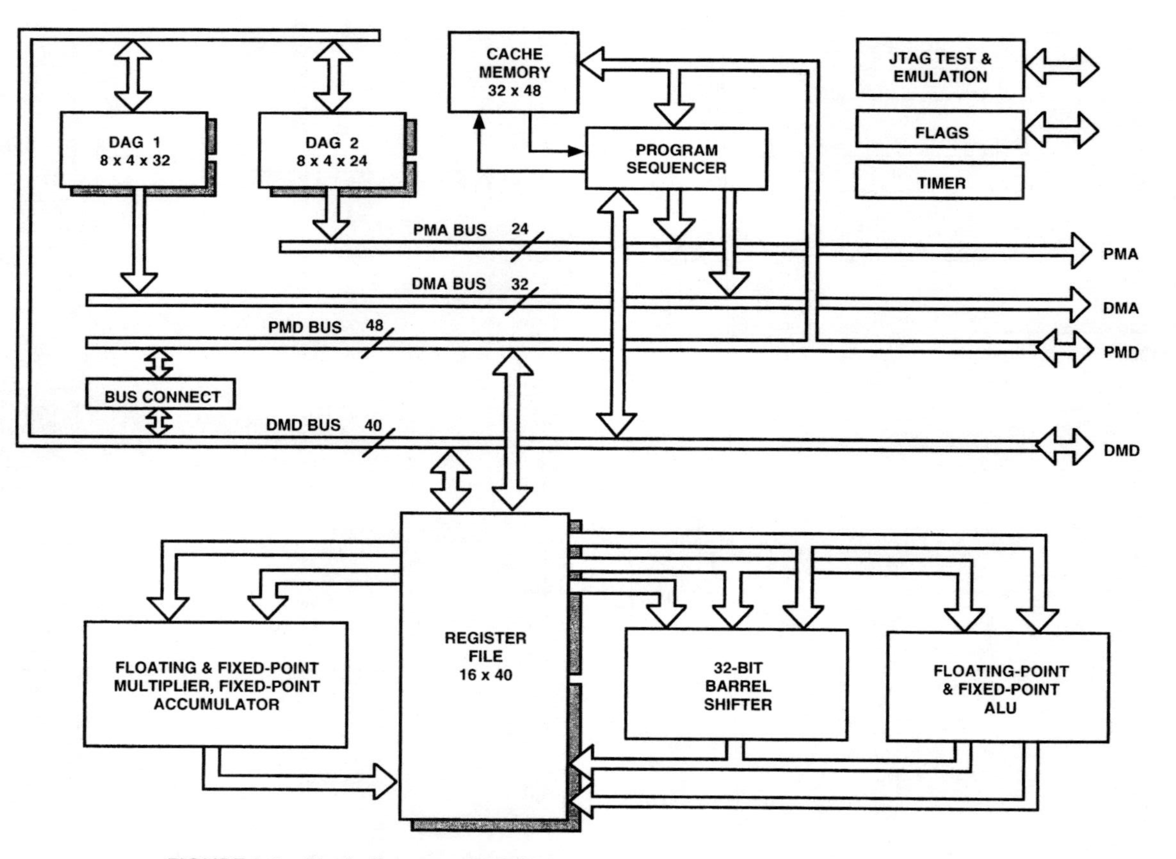

FIGURE 3.3 Block diagram of ADSP-21020 processor (Courtesy Analog Devices.)

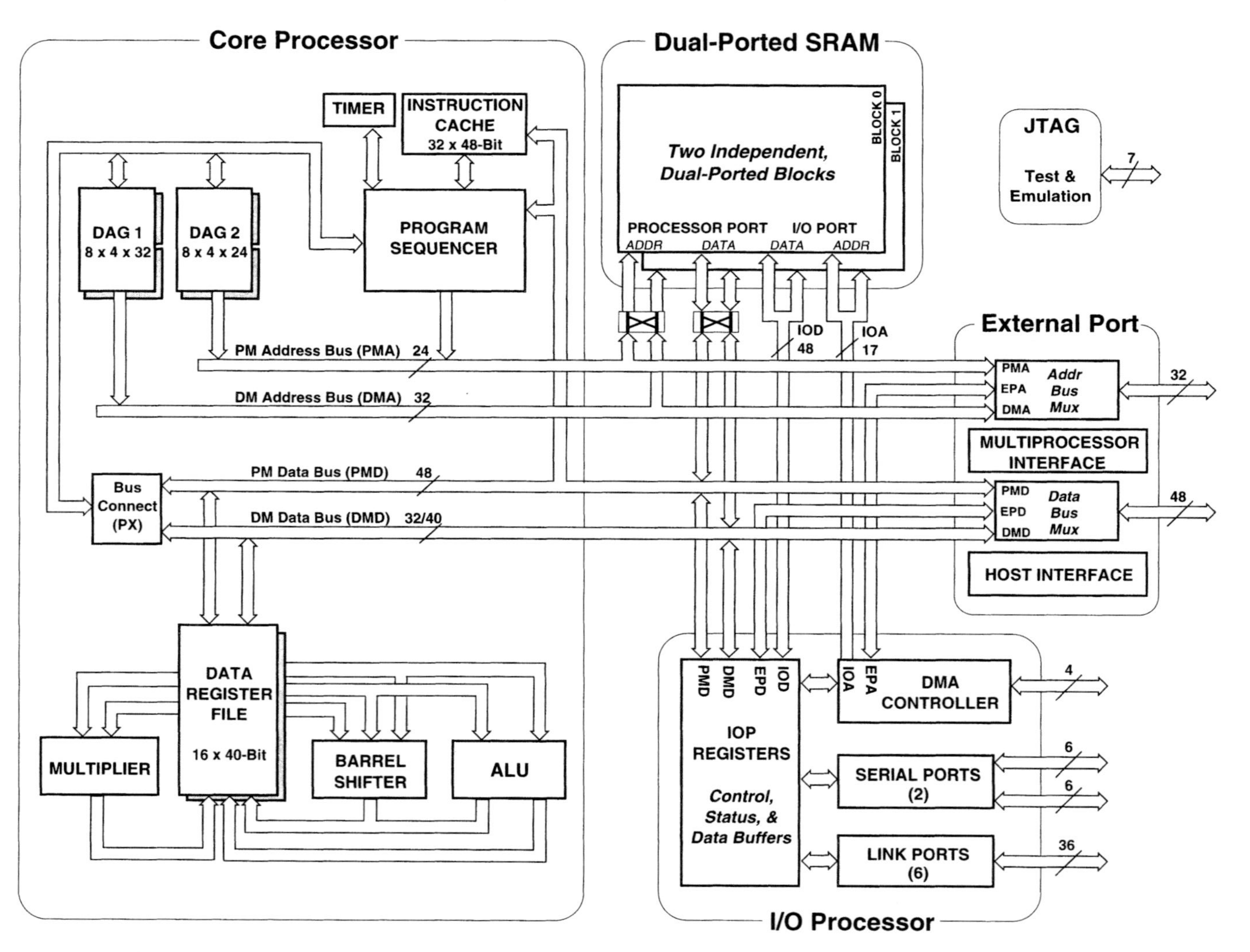

FIGURE 3.4 Block diagram of ADSP-21060 processor (Courtesy Analog Devices.)

with the core processor. The following is a brief description of these processors provided by Analog Devices.

The ADSP-210XX processors provide fast, flexible arithmetic computation units, unconstrained data flow to and from the computation units, extended precision and dynamic range in the computation units, dual address generators, and efficient program sequencing. All instructions execute in a single cycle. It provides one of the fastest cycle times available and the most complete set of arithmetic operations, including seed 1/x, min, max, clip, shift and rotate, in addition to the traditional multiplication, addition, subtraction, and combined addition/subtraction. It is IEEE floating-point compatible and allows interrupts to be generated by arithmetic exceptions or latched status exception handling.

The ADSP-210XX has a modified Harvard architecture combined with a 10-port data register file. In every cycle two operands can be read or written to or from the register file, two operands can be supplied to the ALU, two operands can be supplied to the multiplier, and two results can be received from the ALU and multiplier. The processor's 48-bit orthogonal instruction word supports fully parallel data transfer and arithmetic operations in the same instruction.

The processor handles 32-bit IEEE floating-point format as well as 32-bit integer and fractional formats. It also handles extended precision 40-bit IEEE floating-point formats and carries extended precision throughout its computation units, limiting data truncation errors.

The processor has two data address generators (DAGs) that provide immediate or indirect (pre- and post-modify) addressing. Modulus and bit-reverse addressing operations are supported with no constraints on circular data buffer placement. In addition to zero-overhead loops, the ADSP-210XX supports single-cycle setup and exit for loops. Loops are both nestable (six levels in hardware) and interruptable. The processor supports both delayed and nondelayed branches. In summary, some of the key features of the ADSP-210XX core processor follow:

- 48-bit instruction, 32/40-bit data words
- 80-bit MAC accumulator
- 3-stage pipeline, 63 instruction types
- 32×48-bit instruction cache
- 10-port, 32×40-bit register file (16 registers per set, 2 sets)
- 6-level loop stack
- 24-bit program, 32-bit data address spaces, memory buses
- 1 instruction/cycle (pipelined)
- 1-cycle multiply (32-bit or 40-bit floating-point or 32-bit fixed-point)
- 6-cycle divide (32-bit or 40-bit floating-point)
- 2-cycle branch delay
- Zero overhead loops
- Barrel shifter

- Algebraic syntax assembly language
- Multifunction instructions with 4 operations per cycle
- Dual address generators
- 4-cycle maximum interrupt latency

3.1.3 Texas Instruments TMS320C3X and TMS320C40

Figure 3.5 shows a block diagram of the TMS320C30 microprocessor and Figure 3.6 shows the TMS320C40, both manufactured by Texas Instruments (Houston, TX). The TMS320C30 and TMS320C40, processors are similar in architecture except that the TMS320C40 provides hardware support for multiprocessor configurations. The following is a brief description of the TMS320C30 processor as provided by Texas Instruments.

The TMS320C30 can perform parallel multiply and ALU operations on integer or floating-point data in a single cycle. The processor also possesses a general-purpose register file, program cache, dedicated auxiliary register arithmetic units (ARAU), internal dual-access memories, one DMA channel supporting concurrent I/O, and a short machine-cycle time. High performance and ease of use are products of these features.

General-purpose applications are greatly enhanced by the large address space, multiprocessor interface, internally and externally generated wait states, two external interface ports, two timers, two serial ports, and multiple interrupt structure. High-level language is more easily implemented through a register-based architecture, large address space, powerful addressing modes, flexible instruction set, and well-supported floating-point arithmetic. Some key features of the TMS320C30 are listed below.

- 4 stage pipeline, 113 instructions
- One 4K × 32-bit single-cycle dual access on-chip ROM block
- Two 1K × 32-bit single-cycle dual access on-chip RAM blocks
- 64 × 32-bit instruction cache
- 32-bit instruction and data words, 24-bit addresses
- 40/32-bit floating-point/integer multiplier and ALU
- 32-bit barrel shifter
- Multiport register file: Eight 40-bit extended precision registers (accumulators)
- Two address generators with eight auxiliary registers and two arithmetic units
- On-chip direct memory access (DMA) controller for concurrent I/O
- Integer, floating-point, and logical operations
- Two- and three-operand instructions
- Parallel ALU and multiplier instructions in a single cycle
- Block repeat capability
- Zero-overhead loops with single-cycle branches
- Conditional calls and returns

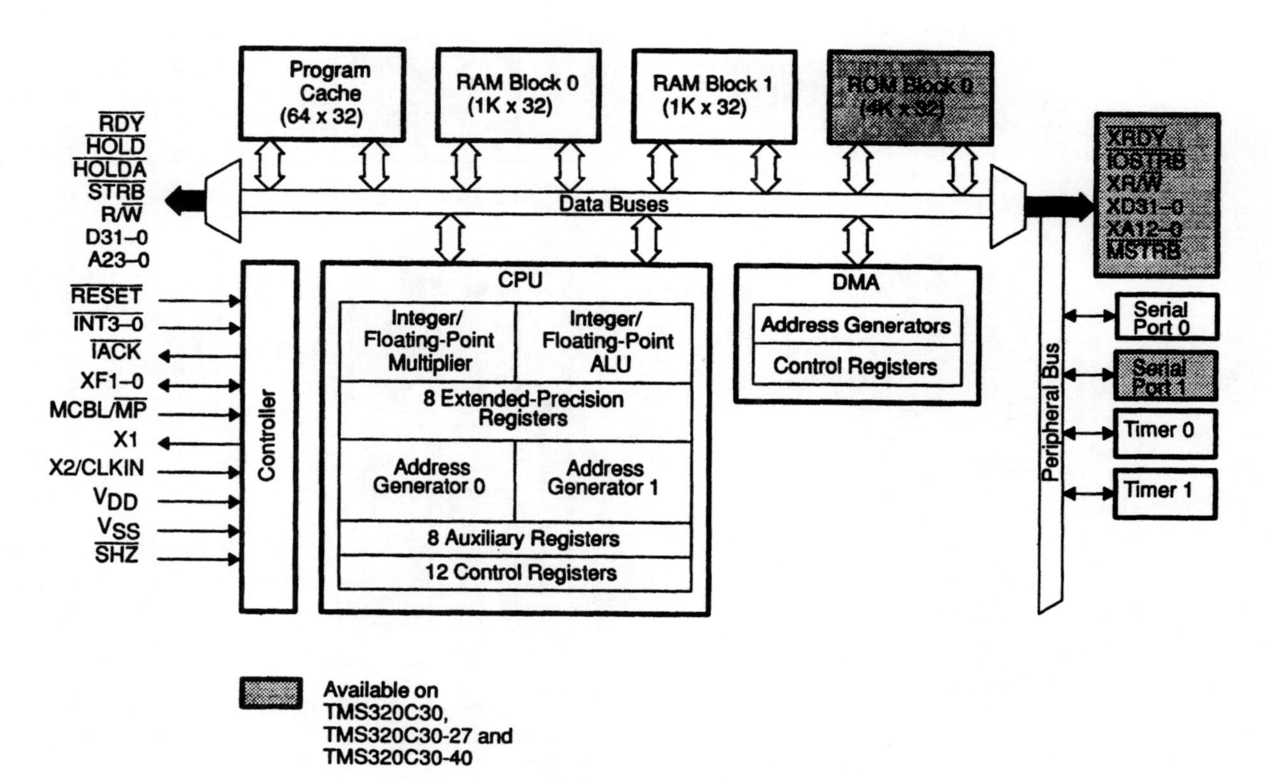

FIGURE 3.5 Block diagram of TMS320C30 and TMS32C31 processor (Courtesy Texas Instruments).

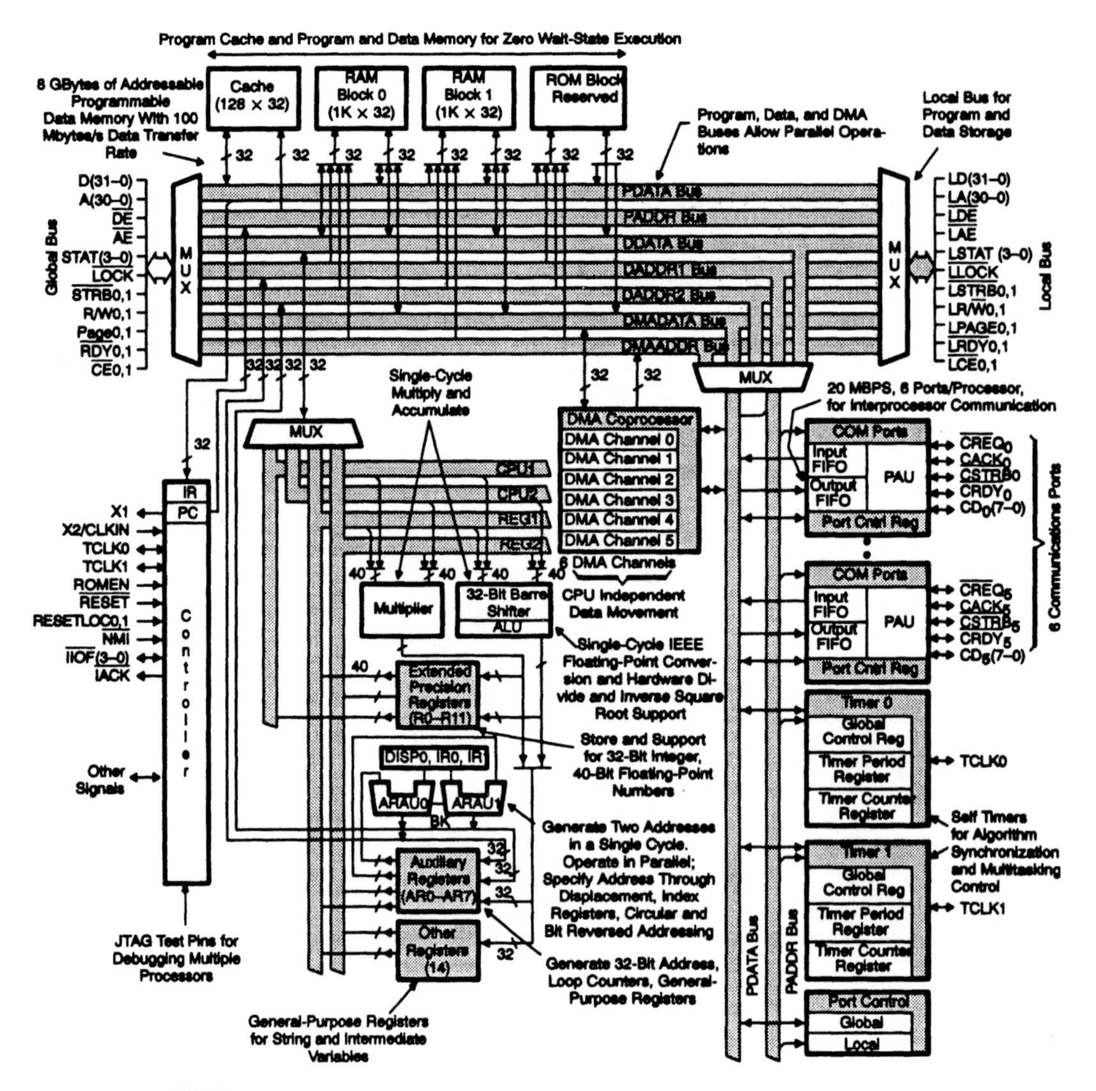

FIGURE 3.6 Block diagram of TMS320C40 processor (Courtesy Texas Instruments).

- Interlocked instructions for multiprocessing support
- Two 32-bit data buses (24- and 13-bit address)
- Two serial ports
- DMA controller
- Two 32-bit timers

3.2 TYPICAL PROGRAMMING TOOLS FOR DSP

The manufacturers of DSP microprocessors typically provide a set of software tools designed to enable the user to develop efficient DSP algorithms for their particular processors. The basic software tools provided include an assembler, linker, C compiler, and simulator. The simulator can be used to determine the detailed timing of an algorithm and then optimize the memory and register accesses. The C compilers for DSP processors will usually generate assembly source code so that the user can see what instructions are generated by the compiler for each line of C source code. The assembly code can then be optimized by the user and then fed into the assembler and linker.

Most DSP C compilers provide a method to add in-line assembly language routines to C programs (see section 3.3.2). This allows the programmer to write highly efficient assembly code for time-critical sections of a program. For example, the autocorrelation function of a sequence may be calculated using a function similar to a FIR filter where the coefficients and the data are the input sequence. Each multiply-accumulate in this algorithm can often be calculated in one cycle on a DSP microprocessor. The same C algorithm may take 4 or more cycles per multiple-accumulate. If the autocorrelation calculation requires 90 percent of the time in a C program, then the speed of the program can be improved by a factor of about 3 if the autocorrelation portion is coded in assembly language and interfaced to the C program (this assumes that the assembly code is 4 times faster than the C source code). The amount of effort required by the programmer to create efficient assembly code for just the autocorrelation function is much less than the effort required to write the entire program in assembly language.

Many DSP software tools come with a library of DSP functions that provide highly optimized assembly code for typical DSP functions such as FFTs and DFTs, FIR and IIR filters, matrix operations, correlations, and adaptive filters. In addition, third parties may provide additional functions not provided by the manufacturer. Much of the DSP library code can be used directly or with small modifications in C programs.

3.2.1 Basic C Compiler Tools

AT&T DSP32C software development tools. The DSP32C's C Compiler provides a programmer with a readable, quick, and portable code generation tool combined with the efficiency provided by the assembly interface and extensive set of library routines. The package provides for compilation of C source code into DSP32 and

DSP32C assembly code, an assembler, a simulator, and a number of other useful utilities for source and object code management. The three forms of provided libraries are:

- libc A subset of the Standard C Library
- libm Math Library
- libap Application Software Library, complete C-callable set of DSP routines.

DSP32C support software library. This package provides assembly-level programming. Primary tools are the assembler, linker/loader, a make utility that provides better control over the assembly and link/load task, and a simulator for program debugging. Other utilities are: library archiver, mask ROM formatter, object file dumper, symbol table lister, object file code size reporter, and EPROM programmer formatter. The SL package is necessary for interface control of AT&T DSP32C Development Systems.

The Application Library has over seven dozen subroutines for arithmetic, matrix, filter, adaptive filter, FFT, and graphics/imaging applications. All files are assembly source and each subroutine has an example test program. Version 2.2.1 adds four routines for sample rate conversion.

AT&T DSP3210 software development tools. This package includes a C language compiler, libraries of standard C functions, math functions, and digital signal processing application functions. A C code usage example is provided for each of the math and application library routines. The C Compiler also includes all of the assembler, simulator, and utility programs found in the DSP3210 ST package. Since the C libraries are only distributed as assembled and then archived ".a" files, a customer may also find the DSP3210-AL package useful as a collection of commented assembly code examples.

DSP3210 support software library. The ST package provides assembly level programming. The primary tools of the package are the assembler, linker/loader, and a simulator for program development, testing, and debugging. A 32C to 3210 assembly code translator assists developers who are migrating from the DSP32C device. Additional utilities are library archiver, mask ROM formatter, object code disassembler, object file dumper, symbol table lister, and object code size reporter. The AT&T Application Software Library includes over ninety subroutines of typical operations for arithmetic, matrix, filter, adaptive filter, FFT, and graphics/imaging applications. All files are assembly source and each subroutine has an example test program.

Analog devices ADSP-210XX C tools. The C tools for the ADSP-21000 family let system developers program ADSP-210XX digital signal processors in ANSI C. Included are the following tools: the G21K C compiler, a runtime library of C functions, and the CBUG C Source-Level Debugger. G21K is Analog Devices' port of GCC, the GNU C compiler from the Free Software Foundation, for the ADSP-21000 family of digital signal processors. G21K includes Numerical C, Analog Devices' numerical processing extensions to the C language based on the work of the ANSI Numerical C Extensions Group (NCEG) subcommittee.

The C runtime library functions perform floating-point mathematics, digital signal processing, and standard C operations. The functions are hand-coded in assembly language for optimum runtime efficiency. The C tools augment the ADSP-21000 family assembler tools, which include the assembler, linker, librarian, simulator, and PROM splitter.

Texas Instruments TMS320C30 C tools. The TMS320 floating-point C compiler is a full-featured optimizing compiler that translates standard ANSI C programs into TMS320C3x/C4x assembly language source. The compiler uses a sophisticated optimization pass that employs several advanced techniques for generating efficient, compact code from C source. General optimizations can be applied to any C code, and target-specific optimizations take advantage of the particular features of the TMS320C3x/C4x architecture. The compiler package comes with two complete runtime libraries plus the source library. The compiler supports two memory models. The small memory model enables the compiler to efficiently access memory by restricting the global data space to a single 64K-word data page. The big memory model allows unlimited space.

The compiler has straightforward calling conventions, allowing the programmer to easily write assembly and C functions that call each other. The C preprocessor is integrated with the parser, allowing for faster compilation. The Common Object File Format (COFF) allows the programmer to define the system's memory map at link time. This maximizes performance by enabling the programmer to link C code and data objects into specific memory areas. COFF also provides rich support for source-level debugging. The compiler package includes a utility that interlists original C source statements into the assembly language output of the compiler. This utility provides an easy method for inspecting the assembly code generated for each C statement.

All data sizes (**`char`**, **`short`**, **`int`**, **`long`**, **`float`**, and **`double`**) are 32 bits. This allows all types of data to take full advantage of the TMS320Cx/C4x's 32-bit integer and floating-point capabilities. For stand-alone embedded applications, the compiler enables linking all code and initialization data into ROM, allowing C code to run from reset.

3.2.2 Memory Map and Memory Bandwidth Considerations

Most DSPs use a Harvard architecture with three memory buses (program and two data memory paths) or a modified Harvard architecture with two memory buses (one bus is shared between program and data) in order to make filter and FFT algorithms execute much faster than standard von Neumann microprocessors. Two separate data and address busses allow access to filter coefficients and input data in the same cycle. In addition, most DSPs perform multiply and addition operations in the same cycle. Thus, DSPs execute FIR filter algorithms at least four times faster than a typical microprocessor with the same MIPS rating.

The use of a Harvard architecture in DSPs causes some difficulties in writing C programs that utilize the full potential of the multiply-accumulate structure and the multiple memory busses. All three manufacturers of DSPs described here provide a method to assign separate physical memory blocks to different C variable types. For example, auto variables that are stored on the heap can be moved from internal memory to external memory by assigning a different address range to the heap memory segment. In the assembly language generated by the compiler the segment name for a particular C variable or array can be changed to locate it in internal memory for faster access or to allow it to be accessed at the same time as the other operands for the multiply or accumulate operation. Memory maps and segment names are used by the C compilers to separate different types of data and improve the memory bus utilization. Internal memory is often used for coefficients (because there are usually fewer coefficients) and external memory is used for large data arrays.

The ADSP-210XX C compiler also supports special keywords so that any C variable or array can be placed in program memory or data memory. The program memory is used to store the program instructions and can also store floating-point or integer data. When the processor executes instructions in a loop, an instruction cache is used to allow the data in program memory (PM) and data in the data memory (DM) to flow into the ALU at full speed. The **pm** keyword places the variable or array in program memory, and the **dm** keyword places the variable or array in data memory. The default for static or global variables is to place them in data memory.

3.2.3 Assembly Language Simulators and Emulators

Simulators for a particular DSP allow the user to determine the performance of a DSP algorithm on a specific target processor before purchasing any hardware or making a major investment in software for a particular system design. Most DSP simulator software is available for the IBM-PC, making it easy and inexpensive to evaluate and compare the performance of several different processors. In fact, it is possible to write all the DSP application software for a particular processor before designing or purchasing any hardware. Simulators often provide profiling capabilities that allow the user to determine the amount of time spent in one portion of a program relative to another. One way of doing this is for the simulator to keep a count of how many times the instruction at each address in a program is executed.

Emulators allow breakpoints to be set at a particular point in a program to examine registers and memory locations, to determine the results from real-time inputs. Before a breakpoint is reached, the DSP algorithm is running at full speed as if the emulator were not present. An in-circuit emulator (ICE) allows the final hardware to be tested at full speed by connecting to the user's processor in the user's real-time environment. Cycle counts can be determined between breakpoints and the hardware and software timing of a system can be examined.

Emulators speed up the development process by allowing the DSP algorithm to run at full speed in a real-time environment. Because simulators typically execute DSP programs several hundred times slower than in real-time, the wait for a program to reach a particular breakpoint in a simulator can be a long one. Real world signals from A/D converters can only be recorded and then later fed into a simulator as test data. Although the test data may test the algorithm performance (if enough test data is available), the timing of the algorithm under all possible input conditions cannot be tested using a simulator. Thus, in many real-time environments an emulator is required.

The AT&T DSP32C simulator is a line-oriented simulator that allows the user to examine all of the registers and pipelines in the processor at any cycle so that small programs can be optimized before real-time constraints are imposed. A typical computer dialog (user input is shown in bold) using the DSP32C simulator is shown below (courtesy of AT&T):

```
$im: SHOWRW=1
$im: b end
bp set at addr 0x44
$im: run
12 | r000004* _______* _______* _______* |0000: r11 = 0x7f(127)
16 | r000008* _______* _______* w00007c* |0004: * r2 = r111
20 | r00000c**_______* _______* r00007c* |0008: a3 = *r2
25 | r000010**_______* r5a5a5a* _______* |000c: r101 = * r1
30 | r000014* _______* _______* _______* |0010: NOP
34 | r000018* _______* _______* _______* |0014: r10 = r10 + 0xff81(-127)
38 | r00001c* _______* _______* w000080* |0018: * r3 = r10
42 | r000020**_______* _______* r000080* |001c: *r3 = a0 = float(*r3)
47 | r000024**_______* _______* _______* |0020: a0 = a3 * a3
52 | r000028* _______* r000074* r000070**|0024: a1 = *r4- + a3 * *r4-
57 | r00002c**_______* r000068* r000064**|0028: a2 = *r5- + a3 * *r5-
63 | r000030**w000080**_______* _______* |002c: a0 = a0 * a3
69 | r000034* _______* _______* r00006c* |0030: a1 = *r4 + a1 * a3
73 | r000038**_______* r00005c* r000058**|0034: a3 = *r6- + a3 * *r6-
79 | r00003c**_______* r00007c* r000060**|0038: a2 = *r5 + a2 * *r2
85 | r000040**_______* _______* _______* |003c: a1 = a1 + a0
90 | r000044* _______* r000080* _______* |0040: *r7 = a0 = a1 + *r3
breakpoint at end{0x000044} decode:*r7 = a0 = a1 + *r3
$im: r7.f
r7 = 16.000000
$im: nwait.d
nwait = 16
```

In the above dialog the flow of data in the four different phases of the DSP32C instruction cycle are shown along with the assembly language operation being performed. The cycle count is shown on the left side. Register r7 is displayed in floating-point after the breakpoint is reached and the number of wait states is also displayed. Memory reads are indicated with an **r** and memory writes with a **w**. Wait states occurring when the same memory is used in two consecutive cycles are shown with ******.

AT&T DSP32C EMULATION SYSTEM DESCRIPTION

- Real-time application development
- Interactive interface to the ICE card(s) provided by the DSP32C simulator through the high-speed parallel interface on the PC Bus Interface Card

- Simulator-controlled software breakpoints; examine and change memory contents or device registers
- For multi-ICE configuration selection of which ICE card is active by the simulator

Figure 3.7 shows a typical screen from the ADSP-21020 screen-oriented simulator. This simulator allows the timing and debug of assembly language routines. The user interface of the simulator lets you completely control program execution and easily change the contents of registers and memory. Breakpoints can also be set to stop execution of the program at a particular point. The ADSP-21020 simulator creates a representation of the device and memory spaces as defined in a system architecture file. It can also simulate input and output from I/O devices using simulated data files. Program execution can be observed on a cycle-by-cycle basis and changes can be made on-line to correct errors. The same screen based interface is used for both the simulator and the in-circuit emulator. A C-source level debugger is also available with this same type of interface (see section 3.3.1).

Figure 3.8 shows a typical screen from the TMS320C30 screen-oriented simulator. This simulator allows the timing of assembly language routines as well as C source code, because it shows the assembly language and C source code at the same time. All of the registers can be displayed after each step of the processor. Breakpoints can also be set to stop execution of the program at a particular point.

```
File  Core  Memory  Execution  Setup  Help
   Data Memory (Fixed)
_play_fifo:                                  Program Memory (Disassembled/T)
RAM   [00000000] -107811072          RAM    [0002de] comp(r2,r4);
RAM   [00000001] 147128993           RAM    [0002df] if ge jump _L20 (db);
RAM   [00000002] 156633879           RAM    [0002e0] nop;
RAM   [00000003] -68615185           RAM    [0002e1] nop;
RAM   [00000004] -372968788          RAM    [0002e2] r2=dm(0xfffffffa,i6);
RAM   [00000005] -569122157          RAM   >[0002e3] r4=ashift r2 by 0x3;
                                     RAM    [0002e4] r9=0x32f5;
 Cycle Counter                       RAM    [0002e5] r2=r4+r9;
    578324791                        RAM    [0002e6] r12=0x473f;
                                     RAM    [0002e7] r8=0x4749;
 Active Register File (Hex)          RAM    [0002e8] r4=r2;
R0: 458dde5a00  R8:  0000000000      RAM    [0002e9] i13=0x2ec;
R1: 0000000100  R9:  0000000000      RAM  * [0002ea] jump _note (db);
R2: 0000000000  R10: 0000000000      RAM    [0002eb] r2=i6;
R3: 7db1aab200  R11: ffffffd00       RAM    [0002ec] i6=i7;
R4: 0000000400  R12: 458dde5a00      RAM    [0002ed] r12=r0;
R5: 457d200000  R13: c833101cdb      RAM    [0002ee] r8=dm(0xfffffff5,i6);
R6: 0000000000  R14: ffffffffff      RAM    [0002ef] f2=f8+f12;
R7: 8a01018056  R15: 8fc3f0f8ff      RAM    [0002f0] dm(0xfffffff5,i6)=r2;
                                     _L21:
Target Halted                                                        08:12:47
```

FIGURE 3.7 ADSP-21020 simulator displaying assembly language and processor registers (Courtesy Analog Devices.)

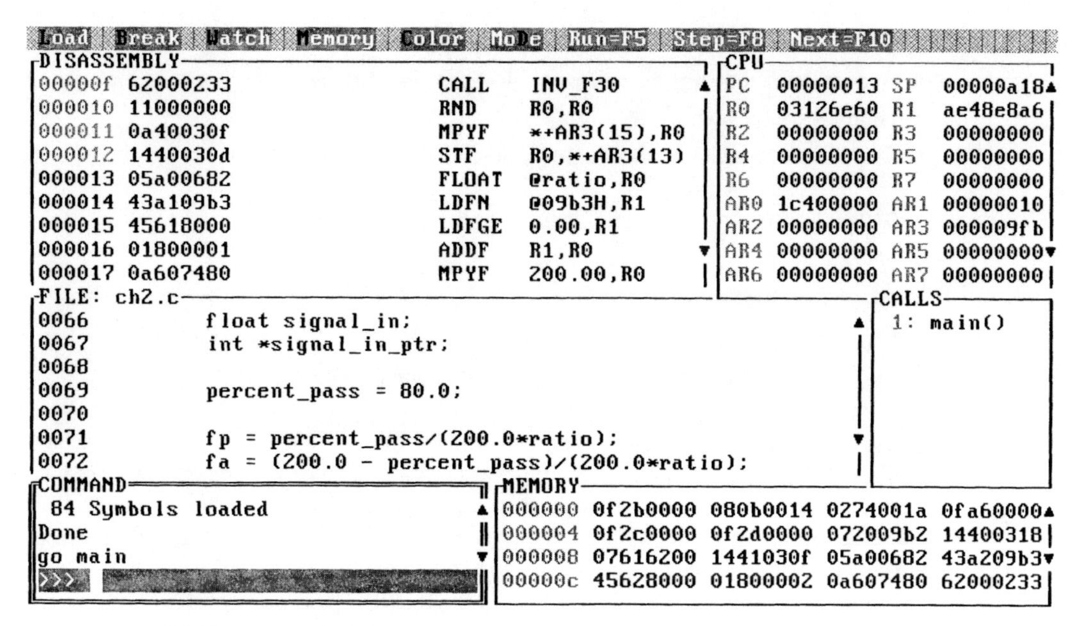

FIGURE 3.8 TMS320C30 simulator in mixed mode assembly language mode (Courtesy Texas Instruments.)

3.3 ADVANCED C SOFTWARE TOOLS FOR DSP

This section describes some of the more advanced software tools available for floating-point DSP microprocessors. Source-level debugging of C source code is described in the next section. Section 3.3.2 describes several assembly language interfaces often used in DSP programming to accelerate key portions of a DSP algorithm. Section 3.3.3 illustrates the numeric C extensions to the C language using DSP algorithms as examples (see Section 2.10 for a description of numeric C).

3.3.1 Source Level Debuggers

Communication Automation & Control (CAC), Inc. (Allentown, PA) offers a debugger for DSP32C assembly language with C-source debugging capability. Both versions are compatible with the following vendors' DSP32C board for the AT computer under MS-DOS: all CAC boards, Ariel, AT&T DSP32C-DS and ICE, Burr-Brown ZPB34, Data Translation, Loughborough Sound Images, and Surrey Medical Imaging Systems. C-source code of the drivers is provided to enable the user to port either debugger to an unsupported DSP32C based board.

Both D3EMU (assembly language only) and D3BUG (C-source and mixed assembly code) are screen-oriented user-friendly symbolic debuggers and have the following features:

- Single-step, multiple breakpoints, run at full speed
- Accumulator/register/mem displays updated automatically after each step
- Global variables with scrolling
- Stand-alone operation or callable by host application

D3BUG ONLY

- C-source and assembly language mixed listing display modes.
- Local and watch variables.
- Stack trace display.
- Multiple source files/directories with time stamp check.

Figure 3.9 shows a typical screen generated using the D3BUG source level debugger with the DSP32C hardware executing the program. Figure 3.9 shows the mixed assembly-C source mode of operation with the DSP32C registers displayed. Figure 3.10 shows the C source mode with the global memory location displayed as the entire C program is executed one C source line at a time in this mode.

Figure 3.11 shows a typical screen from the ADSP-21020 simulator when C source level debugging is being performed using CBUG. C language variables can be displayed and the entire C program can be executed one C source line at a time in this mode. This same type of C source level debug can also be performed using the in-circuit emulator.

```
acc break cont disk goto halt i/o mem code quit reg step vars mix !-DOS ?-help
                                                         REGISTERS
0000b4: 942effe8   r1e=r14+0xffffe8                  1:0x030008 freq2
0000b8: 30000477   *r14++=a0=*r1                     2:0xfff034
0000bc: 00000000   nop                               3:0xfff03c
0000c0: 00000000   nop                               4:0x030800 _1
0000c4: c0610008   r1e=freq2                         5:0x088191
0000c8: 30000477   *r14++=a0=*r1                     6:0xfffd9a
0000cc: 00000000   nop                               7:0xfffffd
                                                     8:0x1cf81b
0000d4: c01400d8   r18e=0x00d8                       9:0xbfa335
0000d8: 9a8e0008   r14e=r14-8                       10:0xfffffc
0000dc: c0610010   r1e=freq_ratio2                  11:0x00405a
0000e0: 30200008   *r1=a1=a0                        12:0xf933e3
0000e4: 00000000   nop                              13:0xfff000
0061> oscinit(freq_ratio1,state_variables1);        14:0xfff038
0000e8: c0610014   r1e=state_variables1             15:0xffffff
0000ec: 1fe101d5   *r14++r19=r1e                    16:0xffffff
0000f0: c061000c   r1e=freq_ratio1                  17:0x5ee7ff
0000f4: 30000477   *r14++=a0=*r1                    18:0x0000a0
0000f8: 00000000   nop                              19:0x000004
0000fc: e01404d4   call oscinit (r18)               20:0x16bf11
000100: c0140104   r18e=0x0104                      21:0x000008
000104: 9a8e0008   r14e=r14-8                       22:0xffffff
a0: 0.0000000e+000  a1: 0.0000000e+000  a2: 0.0000000e+000  a3: 1.7000000e+036
```

FIGURE 3.9 DSP32C debugger D3BUG in mixed assembly C-source mode (Courtesy Communication Automation & Control (CAC), Inc. (Allentown, PA).)

```
acc break cont disk goto halt i/o mem code quit reg step vars mix !-DOS ?-help
0050                                                         ══ GLOBAL VARIABLES ══
0051  /* Select two frequencies */                             data1[0]
0052>  freq1 = 576.0;                                         03002c:  2.685559e-003
0053>  freq2 = 1472.0;                                         data2[0]
0054                                                          03022c:  2.685559e-003
0055  /* Calculate the frequency ratio between the sel         data_in[0]
0056  /*  sampling rate for both oscillators. */              03042c:  2.685559e-003
0057>  freq_ratio1 = freq1/sample_rate;                        data_out[0]
0058>  freq_ratio2 = freq2/sample_rate;                       03062c:  2.685559e-003
0059                                                           errno
0060  /* Initialize each oscillator */                        030000:    805319799
0061>  oscinit(freq_ratio1,state_variables1);                  find_max
0062>  oscinit(freq_ratio2,state_variables2);                 00038c:    536084951
0063                                                           freq1
0064  /* Generate 128 samples for each oscillator */          030004:  1.933384e+026
0065>  oscN(state_variables1,128,data1);                       freq2
0066>  oscN(state_variables2,128,data2);                      030008:  1.933384e+026
0067                                                           freq_ratio1
0068  /* Add the two waveforms together */                    03000c:  1.933384e+026
0069>  add_tones(data1,data2);                                 freq_ratio2
0070                                                          030010:  1.933384e+026
0071  /* Now compute the fft using the AT&T applicatio         log10
0072>  rffta(128,7,data_in);                                  000410:    809501047
0073
```

FIGURE 3.10 DSP32C debugger D3BUG in C-source mode (Courtesy Communication Automation & Control (CAC), Inc. (Allentown, PA).)

```
File  Core  Memory  Execution  Setup  Help
──────────────────────────── CBUG (mu21k.exe) ────────────────────────────
 <Continue>   <Step>   <Next>   <Finish>   <Break>   <Up>   <Down>
 <Execution..>  <Breaks..>  <Data..>  <Context..>  <Symbols..>  <Modes..>
─────────────────────────────── mu21k.c ──────────────────────────────────
    83:          for(i = 0 ; i < endi ; i++) {
    84:              sig_out = 0.0;
    85:              for(v = 0 ; v < vnum ; v++) {
    86:                  sig_out += note(&notes[v],tbreaks,rates);
    87:              }
    88:              sendout(sig_out);
    89:          }
    90:      }                                        ──────C expr──────
    91:                                               >sig_out
    92:      flush();                                 -9813.2988
    93:      flags(0); /* turn off LED */
────────────────────────────── CBUG Status ───────────────────────────────
$ No debug symbols for _key_down().
Stepping into code with symbols.
Err: User halt.  Do a CBUG Step/Next to resume C debugging.
$ Err: User halt.  Do a CBUG Step/Next to resume C debugging.

Target Halted                                                    08:16:49
```

FIGURE 3.11 ADSP-21020 simulator displaying C source code (Courtesy Analog Devices.)

```
 Load  Break  Watch  Memory  Color  Mode  Run=F5  Step=F8  Next=F10
FILE: ch2.c
0070
0071          fp = percent_pass/(200.0*ratio);
0072          fa = (200.0 - percent_pass)/(200.0*ratio);
0073          deltaf = fa-fp;
0074
0075          nfilt = filter_length( att, deltaf, &beta );
0076
0077          lsize = nfilt/ratio;
0078
0079          nfilt = lsize*ratio + 1;
0080 BP>      npair = (nfilt - 1)/2;
0081
0082          for(i = 0 ; i < ratio ; i++) {
0083              h[i] = (float *) calloc(lsize,sizeof(float));
0084              if(!h[i]) {
COMMAND                                                  CALLS
                              WATCH                       1: main()
Loading ch2.out                1: i 3
 84 Symbols loaded             2: clk 1906
Done
go main
>>>
```

FIGURE 3.12 TMS320C30 simulator in C-source mode (Courtesy Texas Instruments.)

Figure 3.12 shows a typical screen from the TMS320C30 simulator when C source level debugging is being performed. C language variables can be displayed and the entire C program can be executed one C source line at a time in this mode.

3.3.2 Assembly-C Language Interfaces

The DSP32C/DSP3210 compiler provides a macro capability for in-line assembly language and the ability to link assembly language functions to C programs. In-line assembly is useful to control registers in the processor directly or to improve the efficiency of key portions of a C function. C variables can also be accessed using the optional operands as the following scale and clip macro illustrates:

```
asm void scale(flt_ptr,scale_f,clip)
{
% ureg flt_ptr,scale_f,clip;
a0 = scale_f * *flt_ptr
a1 = -a0 + clip
a0 = ifalt(clip)
*flt_ptr++ = a0 = a0
}
```

Assembly language functions can be easily linked to C programs using several macros supplied with the DSP32C/DSP3210 compiler that define the beginning and the end of the assembly function so that it conforms to the register usage of the C compiler.

The macro **@B** saves the calling function's frame pointer and the return address. The macro **@E0** reads the return address off the stack, performs the stack and frame pointer adjustments, and returns to the calling function. The macros do not save registers used in the assembly language code that may also be used by the C compiler—these must be saved and restored by the assembly code. All parameters are passed to the assembly language routine on the stack and can be read off the stack using the macro **param()**, which gives the address of the parameter being passed.

The ADSP-210XX compiler provides an **asm()** construct for in-line assembly language and the ability to link assembly language functions to C programs. In-line assembly is useful for directly accessing registers in the processor, or for improving the efficiency of key portions of a C function. The assembly language generated by **asm()** is embedded in the assembly language generated by the C compiler. For example, **asm("bit set imask 0x40;")** will enable one of the interrupts in one cycle. C variables can also be accessed using the optional operands as follows:

```
asm("%0=clip %1 by %2;" : "=d" (result) : "d" (x), "d" (y));
```

where **result**, **x** and **y** are C language variables defined in the C function where the macro is used. Note that these variables will be forced to reside in registers for maximum efficiency.

ADSP-210XX assembly language functions can be easily linked to C programs using several macros that define the beginning and end of the assembly function so that it conforms to the register usage of the C compiler. The macro **entry** saves the calling function's frame pointer and the return address. The macro **exit** reads the return address off the stack, performs the stack and frame pointer adjustments, and returns to the calling function. The macros do not save registers that are used in the assembly language code which may also be used by the C compiler—these must be saved and restored by the assembly code. The first three parameters are passed to the assembly language routine in registers r4, r8, and r12 and the remaining parameters can be read off the stack using the macro **reads()**.

The TMS320C30 compiler provides an **asm()** construct for in-line assembly language. In-line assembly is useful to control registers in the processor directly. The assembly language generated by **asm()** is embedded in the assembly language generated by the C compiler. For example, **asm(" LDI @MASK,IE")** will unmask some of the interrupts controlled by the variable **MASK**. The assembly language routine must save the calling function frame pointer and return address and then restore them before returning to the calling program. Six registers are used to pass arguments to the assembly language routine and the remaining parameters can be read off the stack.

3.3.3 Numeric C Compilers

As discussed in section 2.10, numerical C can provide vector, matrix, and complex operations using fewer lines of code than standard ANSI C. In some cases the compiler may be able to perform better optimization for a particular processor. A complex FIR filter can be implemented in ANSI C as follows:

```
typedef struct {
    float real, imag;
} COMPLEX;

COMPLEX float x[1024],w[1024];
COMPLEX *xc,*wc,out;
xc=x;
wc=w;
out.real = 0.0;
out.imag = 0.0;
for(i = 0 ; i < n ; i++) {
    out.real += xc[i].real*wc[i].real - xc[i].imag*wc[i].imag;
    out.imag += xc[i].real*wc[i].imag + xc[i].imag*wc[i].real;
}
```

The following code segment shows the numeric C implementation of the same complex FIR filter:

```
complex float out,x[1024],w[1024];
{
    iter I = n;
    out=sum(x[I]*w[I]);
}
```

The numeric C code is only five lines versus the ten lines required by the standard C implementation. The numeric C code is more efficient, requiring 14 cycles per filter tap versus 17 in the standard C code.

More complicated algorithms are also more compact and readable. The following code segment shows a standard C implementation of a complex FFT without the bit-reversal step (the output data is bit reversed):

```
void fft_c(int n,COMPLEX *x,COMPLEX *w)
{
    COMPLEX u,temp,tm;
    COMPLEX *xi,*xip,*wptr;
    int i,j,le,windex;

    windex = 1;
    for(le=n/2 ; le > 0 ; le/=2) {
        wptr = w;
        for (j = 0 ; j < le ; j++) {
            u = *wptr;
            for (i = j ; i < n ; i = i + 2*le) {
                xi = x + i;
                xip = xi + le;
                temp.real = xi->real + xip->real;
```

```
                    temp.imag = xi->imag + xip->imag;
                    tm.real = xi->real - xip->real;
                    tm.imag = xi->imag - xip->imag;
                    xip->real = tm.real*u.real - tm.imag*u.imag;
                    xip->imag = tm.real*u.imag + tm.imag*u.real;
                    *xi = temp;
                }
                wptr = wptr + windex;
            }
            windex = 2*windex;
        }
    }
```

The following code segment shows the numeric C implementation of a complex FFT without the bit-reversal step:

```
void fft_nc(int n, complex float *x, complex float *w)
{
    int size,sect,deg = 1;
    for(size=n/2 ; size > 0 ; size/=2) {
        for(sect=0 ; sect < n ; sect += 2*size) {
            complex float *x1=x+sect;
            complex float *x2=x1+size;
            { iter I=size;
              for(I) {
                complex float temp;
                temp = x1[I] + x2[I];
                x2[I] = (x1[I] - x2[I]) * w[deg*I];
                x1[I] = temp;
              }
            }
        }
        deg *= 2;
    }
}
```

The twiddle factors (**w**) can be initialized using the following numeric C code:

```
void init_w(int n, complex float *w)
{
    iter I = n;
    float a = 2.0*PI/n;
    w[I] = cosf(I*a) + 1i*sinf(I*a);
}
```

Note that the performance of the **init_w** function is almost identical to a standard C implementation, because most of the execution time is spent inside the cosine and sine functions. The numerical C implementation of the FFT also has an almost identical execution time as the standard C version.

3.4 REAL-TIME SYSTEM DESIGN CONSIDERATIONS

Real-time systems by definition place a hard restriction on the response time to one or more events. In a digital signal processing system the events are usually the arrival of new input samples or the requirement that a new output sample be generated. In this section several real-time DSP design considerations are discussed.

Runtime initialization of constants in a program during the startup phase of a DSP's execution can lead to much faster real-time performance. Consider the pre-calculation of $1/\pi$, which may be used in several places in a particular algorithm. A lengthy divide in each case is replaced by a single multiply if the constant is pre-calculated by the compiler and stored in a static memory area. Filter coefficients can also be calculated during the startup phase of execution and stored for use by the real-time code. The tradeoff that results is between storing the constants in memory which increases the minimum memory size requirements of the embedded system or calculating the constants in real-time. Also, if thousands of coefficients must be calculated, the startup time may become exceeding long and no longer meet the user's expectations. Most DSP software development systems provide a method to generate the code such that the constants can be placed in ROM so that they do not need to be calculated during startup and do not occupy more expensive RAM.

3.4.1 Physical Input/Output (Memory Mapped, Serial, Polled)

Many DSPs provide hardware that supports serial data transfers to and from the processor as well as external memory accesses. In some cases a direct memory access (DMA) controller is also present, which reduces the overhead of the input/output transfer by transferring the data from memory to the slower I/O device in the background of the real-time program. In most cases the processor is required to wait for some number of cycles whenever the processor accesses the same memory where the DMA process is taking place. This is typically a small percentage of the processing time, unless the input or output DMA rate is close to the MIPS rating of the processor.

Serial ports on DSP processors typically run at a maximum of 20 to 30 Mbits/second allowing approximately 2.5 to 4 Mbytes to be transferred each second. If the data input and output are continuous streams of data, this works well with the typical floating-point processor MIPS rating of 12.5 to 40. Only 4 to10 instructions could be executed between each input or output leading to a situation where very little signal processing could be performed.

Parallel memory-mapped data transfers can take place at the MIPs rating of the processor, if the I/O device can accept the data at this rate. This allows for rapid transfers of data in a burst fashion. For example, 1024 input samples could be acquired from a

10 MHz A/D converter at full speed in 100 μsec, and then a FFT power spectrum calculation could be performed for the next 5 msec. Thus, every 5.1 msec the A/D converter's output would be used.

Two different methods are typically used to synchronize the microprocessor with the input or output samples. The first is polling loops and the second is interrupts which are discussed in the next section. Polling loops can be highly efficient when the input and output samples occur at a fixed rate and there are a small number of inputs and outputs. Consider the following example of a single input and single output at the same rate:

```
for(;;) {
    while(*in_status & 1);
    *out = filter(*in)
}
```

It is assumed that the memory addresses of **in**, **out**, and **in_status** have been defined previously as global variables representing the physical addresses of the I/O ports. The data read at **in_status** is bitwise ANDed with 1 to isolate the least significant bit. If this bit is 1, the **while** loop will loop continuously until the bit changes to 0. This bit could be called a "not ready flag" because it indicates that an input sample is not available. As soon as the next line of C code accesses the **in** location, the hardware must set the flag again to indicate that the input sample has been transferred into the processor. After the **filter** function is complete, the returned value is written directly to the output location because the output is assumed to be ready to accept data. If this were not the case, another polling loop could be added to check if the output were ready. The worst case total time involved in the filter function and at least one time through the **while** polling loop must be less than the sampling interval for this program to keep up with the real-time input. While this code is very efficient, it does not allow for any changes in the filter program execution time. If the filter function takes twice as long every 100 samples in order to update its coefficients, the maximum sampling interval will be limited by this larger time. This is unfortunate because the microprocessor will be spending almost half of its time idle in the **while** loop. Interrupt-driven I/O, as discussed in the next section, can be used to better utilize the processor in this case.

3.4.2 Interrupts and Interrupt-Driven I/O

In an interrupt-driven I/O system, the input or output device sends a hardware interrupt to the microprocessor requesting that a sample be provided to the hardware or accepted as input from the hardware. The processor then has a short time to transfer the sample. The interrupt response time is the sum of the interrupt latency of the processor, the time required to save the current context of the program running before the interrupt occurred and the time to perform the input or output operation. These operations are almost always performed in assembly language so that the interrupt response time can be minimized. The advantage of the interrupt-driven method is that the processor is free to perform other tasks, such as processing other interrupts, responding to user requests or slowly changing the parameters associated with the algorithm. The disadvantage of interrupts is the overhead associated with the interrupt latency, context save, and restore associated with the interrupt process.

The following C code example (file INTOUT.C on the enclosed disk) illustrates the functions required to implement one output interrupt driven process that will generate 1000 samples of a sine wave:

```
#include <signal.h>
#include <math.h>
#include "rtdspc.h"

#define SIZE 10

int output_store[SIZE];
int in_inx = 0;
volatile int out_inx = 0;

void sendout(float x);
void output_isr(int ino);

int in_fifo[10000];
int index = 0;

void main()
{
    static float f,a;
    int i,j;
    setup_codec(6);

    for(i = 0 ; i < SIZE-1 ; i++) sendout(i);
    interrupt(SIG_IRQ3, output_isr);

    i = 0;
    j = 1;
    for(;;) {
        for(f=0.0 ; f < 1000.0 ; f += 0.005) {
            sendout(a*sinf(f*PI));
            i += j;
            if(i%25 == 0) {
                a = 100.0*exp(i*5e-5);
                if(a > 30000.0 || i < 0) j = -j;
            }
        }
    }
}
void sendout(float x)
{
```

```
        in_inx++;
        if(in_inx == SIZE) in_inx = 0;
        while(in_inx == out_inx);
        output_store[in_inx] = (int)x;
    }
    void output_isr(int ino)
    {
        volatile int *out = (int *)0x40000002;

        if(index < 10000)
          in_fifo[index++]=16*in_inx+out_inx;

        *out = output_store[out_inx++] << 16;
        if(out_inx == SIZE) out_inx = 0;
    }
```

The C function **output_isr** is shown for illustration purposes only (the code is ADSP-210XX specific), and would usually be written in assembly language for greatest efficiency. The functions **sendout** and **output_isr** form a software first-in first-out (FIFO) sample buffer. After each interrupt the output index is incremented with a circular 0-10 index. Each call to **sendout** increments the **in_inx** variable until it is equal to the **out_inx** variable, at which time the output sample buffer is full and the while loop will continue until the interrupt process causes the **out_inx** to advance. Because the above example generates a new **a** value every 25 samples, the FIFO tends to empty during the **exp** function call. The following table, obtained from measurements of the example program at a 48 KHz sampling rate, illustrates the changes in the number of samples in the software FIFO.

Sample Index	in_inx value	out_inx value	Number of Samples in FIFO
0	2	2	10
1	3	3	10
2	4	4	10
3	4	5	9
4	4	6	8
5	4	7	7
6	4	8	6
7	4	9	5
8	7	0	7
9	9	1	8
10	2	2	10

As shown in the table, the number of samples in the FIFO drops from 10 to 5 and then is quickly increased to 10, at which point the FIFO is again full.

3.4.3 Efficiency of Real-Time Compiled Code

The efficiency of compiled C code varies considerably from one compiler to the next. One way to evaluate the efficiency of a compiler is to try different C constructs, such as **case** statements, nested **if** statements, integer versus floating-point data, **while** loops versus **for** loops and so on. It is also important to reformulate any algorithm or expression to eliminate time-consuming function calls such as calls to exponential, square root, or transcendental functions. The following is a brief list of optimization techniques that can improve the performance of compiled C code.

(1) Use of arithmetic identities—multiplies by 0 or 1 should be eliminated whenever possible especially in loops where one iteration has a multiply by 1 or zero. All divides by a constant can also be changed to multiplies.

(2) Common subexpression elimination—repeated calculations of same subexpression should be avoided especially within loops or between different functions.

(3) Use of intrinsic functions—use macros whenever possible to eliminate the function call overhead and convert the operations to in-line code.

(4) Use of register variables—force the compiler to use registers for variables which can be identified as frequently used within a function or inside a loop.

(5) Loop unrolling—duplicate statements executed in a loop in order to reduce the number of loop iterations and hence the loop overhead. In some cases the loop is completely replaced by in-line code.

(6) Loop jamming or loop fusion—combining two similar loops into one, thus reducing loop overhead.

(7) Use post-incremented pointers to access data in arrays rather than subscripted variables (**x=array[i++]** is slow, **x=*ptr++** is faster).

In order to illustrate the efficiency of C code versus optimized assembly code, the following C code for one output from a 35 tap FIR filter will be used:

```
float in[35],coefs[35],y;
main()
{
  register int i;
  register float *x = in, *w = coefs;
  register float out;

  out = *x++ * *w++;
  for(i = 16 ; i– >= 0; ) {
    out += *x++ * *w++;
    out += *x++ * *w++;
  }
  y=out;
}
```

The FIR C code will execute on the three different processors as follows:

Processor	Optimized C Code Cycles	Optimized Assembly Cycles	Relative Efficiency of C Code (%)
DSP32C	462	187	40.5
ADSP-21020	185	44	23.8
TMS320C30	241	45	18.7

The relative efficiency of the C code is the ratio of the assembly code cycles to the C code cycles. An efficiency of 100 percent would be ideal. Note that this code segment is one of the most efficient loops for the DSP32C compiler but may not be for the other compilers. This is illustrated by the following 35-tap FIR filter code:

```
float in[35],coefs[35],y;
main()
{
  register int i;
  register float *x = in;
  register float *w = coefs;
  register float out;

  out = *x++ * *w++;
  for(i = 0 ; i < 17 ; i++ ) {
    out += *x++ * *w++;
    out += *x++ * *w++;
  }
  y=out;
}
```

This **for**-loop based FIR C code will execute on the three different processors as follows:

Processor	Optimized C Code Cycles	Optimized Assembly Cycles	Relative Efficiency of C Code (%)
DSP32C	530	187	35.3
ADSP-21020	109	44	40.4
TMS320C30	211	45	21.3

Note that the efficiency of the ADSP-21020 processor C code is now almost equal to the efficiency of the DSP32C C code in the previous example.

The complex FFT written in standard C code shown in Section 3.3.3 can be used to illustrate a more complicated algorithm. This C code for a 1024-point complex FFT will execute on the three different processors as follows:

Processor	Optimized C Code Cycles	Optimized Assembly Cycles	Relative Efficiency of C Code (%)
DSP32C	627855	168757	26.9
ADSP-21020	190247	19245	10.1
TMS320C30	322040	40457	12.6

3.4.4 Multiprocessor Architectures

Figure 3.13 shows two simple multiprocessor architectures that could be used to solve the same DSP problem. For a particular DSP application, a set of 1024 input samples must be processed in such a way that 10 output samples are generated for each new set of input samples. The problem is that this must be done in less than 6000 cycles on the target processor (600 cycles/output). A study of the most efficient code reveals that the best that

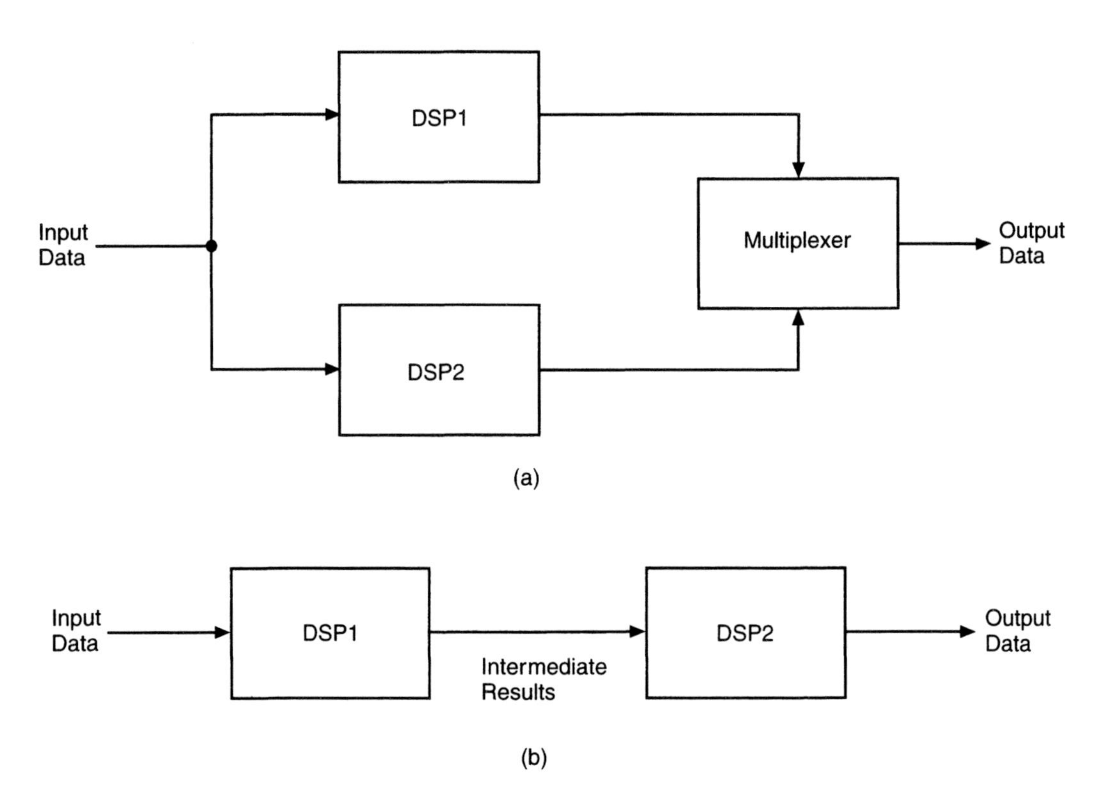

FIGURE 3.13 (a) Block diagram of parallel processor architecture. (b) Block diagram of cascade processor architecture

a single processor can do is 10,000 cycles, not including the time for input and output. Figure 3.13(a) shows the parallel approach where alternating inputs are sent to each processor and the outputs are reassembled by a multiplexer. Figure 3.13(b) shows the cascade approach where an intermediate 512-sample result from the algorithm is sent to the second processor and the final output is generated by the second processor. If it takes one cycle for each input or output on all processors then the two architectures can be compared as follows:

Processor Cycle Usage	**Parallel DSP1**	**Parallel DSP2**	**Cascade DSP1**	**Cascade DSP2**
Input cycles	1024	1024	1024	512
Algorithm cycles	10000	10000	5000	5000
Output cycles	10	10	512	10
Total cycles	11034	11034	6534	5522
Cycles/output	551.7	551.7	—	653.4

In this example, the parallel processor approach meets the goal of 600 cycles per output and the cascade processor approach does not, because the data flow is more optimum in the parallel structure. Note also that the processing time for DSP1 in the cascade approach is what controls the overall performance, since it takes longer to generate the 512 intermediate results. Another advantage of the parallel approach is that both programs in the two DSP processors can be identical, making software development somewhat simpler.

CHAPTER 4

REAL-TIME FILTERING

Filtering is the most commonly used signal processing technique. Filters are usually used to remove or attenuate an undesired portion of a signal's spectrum while enhancing the desired portions of the signal. Often the undesired portion of a signal is random noise with a different frequency content than the desired portion of the signal. Thus, by designing a filter to remove some of the random noise, the signal-to-noise ratio can be improved in some measurable way.

Filtering can be performed using analog circuits with continuous-time analog inputs or using digital circuits with discrete-time digital inputs. In systems where the input signal is digital samples (in music synthesis or digital transmission systems, for example) a digital filter can be used directly. If the input signal is from a sensor which produces an analog voltage or current, then an analog-to-digital converter (A/D converter) is required to create the digital samples. In either case, a digital filter can be used to alter the spectrum of the sampled signal, x_i, in order to produce an enhanced output, y_i. Digital filtering can be performed in either the time domain (see section 4.1) or the frequency domain (see section 4.4), with general-purpose computers using previously stored digital samples or in real-time with dedicated hardware.

4.1 REAL-TIME FIR AND IIR FILTERS

Figure 4.1 shows a typical digital filter structure containing N memory elements used to store the input samples and N memory elements (or delay elements) used to store the output sequence. As a new sample comes in, the contents of each of the input memory elements are copied to the memory elements to the right. As each output sample is formed

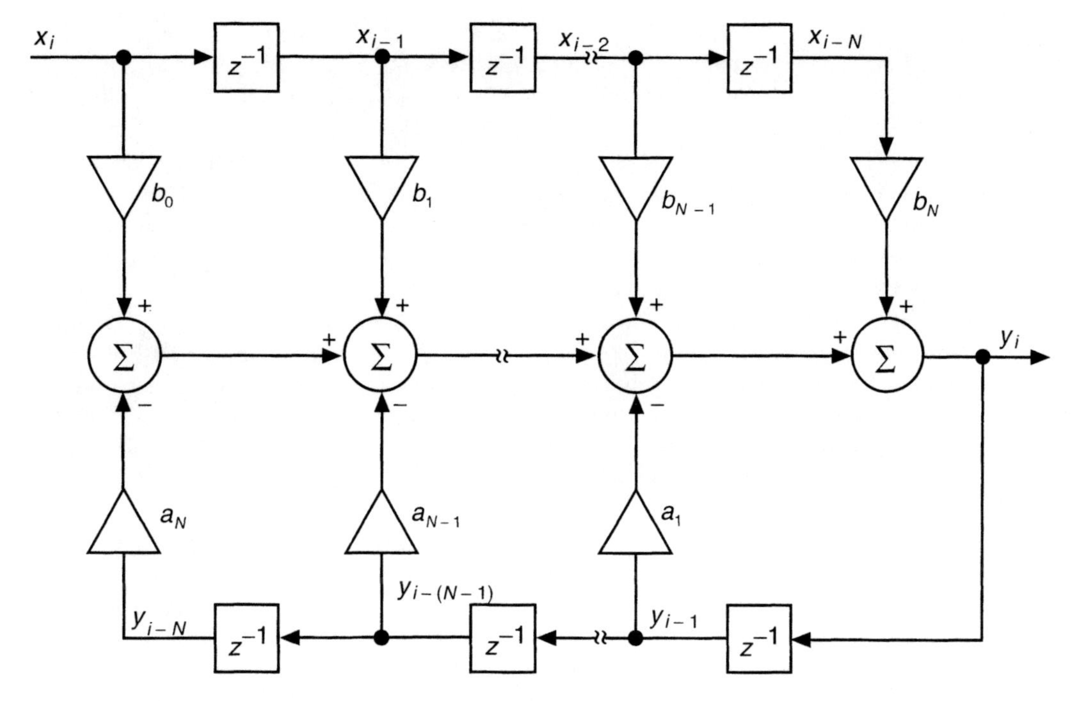

FIGURE 4.1 Filter structure of Nth order filter. The previous N input and output samples stored in the delay elements are used to form the output sum.

by accumulating the products of the coefficients and the stored values, the output memory elements are copied to the left. The series of memory elements forms a digital delay line. The delayed values used to form the filter output are called *taps* because each output makes an intermediate connection along the delay line to provide a particular delay. This filter structure implements the following difference equation:

$$y(n) = \sum_{q=0}^{Q-1} b_q x(n-q) - \sum_{p=1}^{P-1} a_p y(n-p). \tag{4.1}$$

As discussed in Chapter 1, filters can be classified based on the duration of their impulse response. Filters where the a_n terms are zero are called *finite impulse response* (FIR) *filters,* because the response of the filter to an impulse (or any other input signal) cannot change N samples past the last excitation. Filters where one or more of the a_n terms are nonzero are *infinite impulse response* (IIR) *filters.* Because the output of an IIR filter depends on a sum of the N input samples as well as a sum of the past N output samples, the output response is essentially dependent on all past inputs. Thus, the filter output response to any finite length input is infinite in length, giving the IIR filter infinite memory.

Finite impulse response (FIR) *filters* have several properties that make them useful for a wide range of applications. A perfect linear phase response can easily be constructed with an FIR filter allowing a signal to be passed without phase distortion. FIR filters are inherently stable, so stability concerns do not arise in the design or implementation phase of development. Even though FIR filters typically require a large number of multiplies and adds per input sample, they can be implemented using fast convolution with FFT algorithms (see section 4.4.1). Also, FIR structures are simpler and easier to implement with standard fixed-point digital circuits at very high speeds. The only possible disadvantage of FIR filters is that they require more multiplies for a given frequency response when compared to IIR filters and, therefore, often exhibit a longer processing delay for the input to reach the output.

During the past 20 years, many techniques have been developed for the design and implementation of FIR filters. *Windowing* is perhaps the simplest and oldest FIR design technique (see section 4.1.2), but is quite limited in practice. The *window design method* has no independent control over the passband and stopband ripple. Also, filters with unconventional responses, such as *multiple passband filters,* cannot be designed. On the other hand, window design can be done with a pocket calculator and can come close to optimal in some cases.

This section discusses FIR filter design with different equiripple error in the passbands and stopbands. This class of FIR filters is widely used primarily because of the well-known *Remez exchange algorithm* developed for FIR filters by Parks and McClellan. The general Parks-McClellan program can be used to design filters with several passbands and stopbands, digital differentiators, and Hilbert transformers. The FIR coefficients obtained program can be used directly with the structure shown in Figure 4.1 (with the a_n terms equal to zero). The floating-point coefficients obtained can be directly used with floating-point arithmetic (see section 4.1.1).

The Parks-McClellan program is available on the IEEE digital signal processing tape or as part of many of the filter design packages available for personal computers. The program is also printed in several DSP texts (see Elliot, 1987, or Rabiner and Gold, 1975). The program REMEZ.C is a C language implementation of the Parks-McClellan program and is included on the enclosed disk. An example of a filter designed using the REMEZ program is shown at the end of section 4.1.2. A simple method to obtain FIR filter coefficients based on the Kaiser window is also described in section 4.1.2. Although this method is not as flexible as the Remez exchange algorithm it does provide optimal designs without convergence problems or filter length restrictions.

4.1.1 FIR Filter Function

Figure 4.2 shows a block diagram of the FIR real-time filter implemented by the function **`fir_filter`** (shown in Listing 4.1 and contained in the file FILTER.C). The **`fir_filter`** function implements the FIR filter using a history pointer and coefficients passed to the function. The history array is allocated by the calling program and is used to store the previous $N - 1$ input samples (where N is the number of FIR filter coefficients). The last few lines of code in **`fir_filter`** implements the multiplies and accumulates required for the FIR filter of length N. As the history pointer is advanced by using a post-increment, the coefficient pointer is post-decremented. This effectively time reverses the

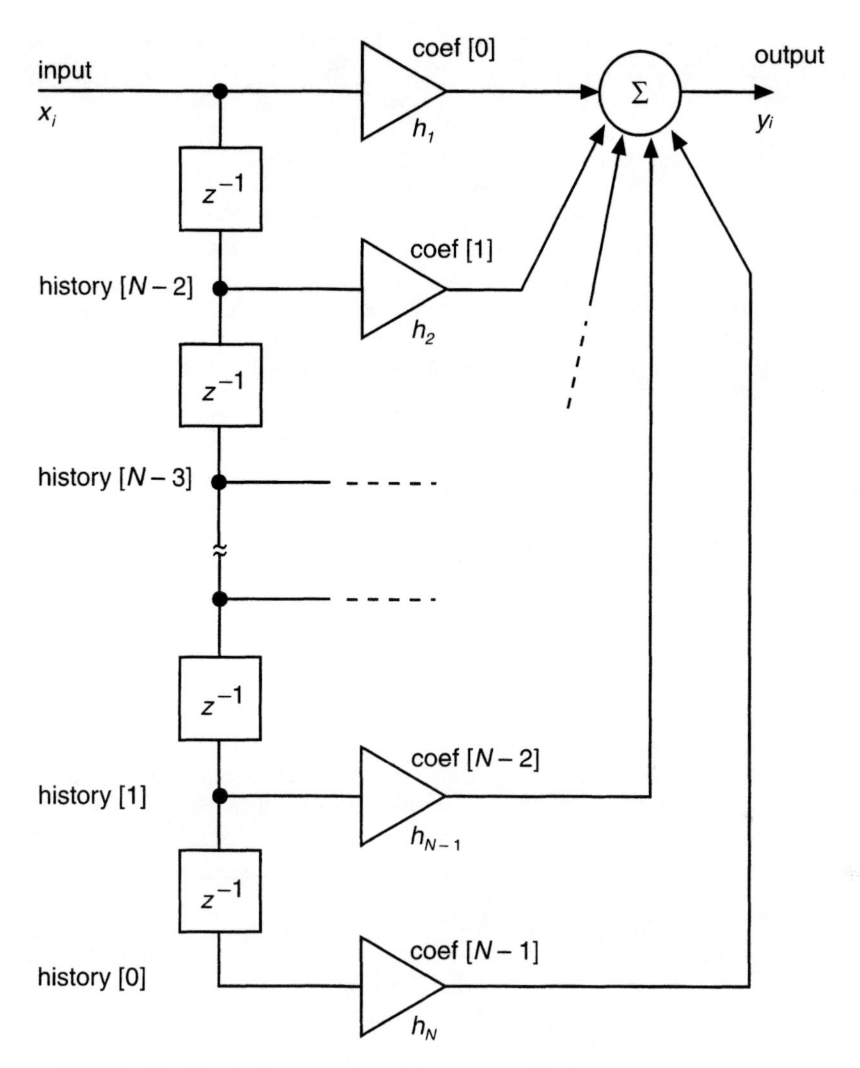

FIGURE 4.2 Block diagram of real-time N tap FIR filter structure as implemented by function **fir_filter**.

coefficients so that a true convolution is implemented. On some microprocessors, post-decrement is not implemented efficiently so this code becomes less efficient. Improved efficiency can be obtained in this case by storing the filter coefficients in time-reversed order. Note that if the coefficients are symmetrical, as for simple linear phase lowpass filters, then the time-reversed order and normal order are identical. After the **for** loop and $N-1$ multiplies have been completed, the history array values are shifted one sample toward **history[0]**, so that the new input sample can be stored in **history[N-1]**. The **fir_filter** implementation uses pointers extensively for maximum efficiency.

```
/**************************************************************************

fir_filter - Perform fir filtering sample by sample on floats

Requires array of filter coefficients and pointer to history.
Returns one output sample for each input sample.

float fir_filter(float input,float *coef,int n,float *history)

    float input        new float input sample
    float *coef        pointer to filter coefficients
    int n              number of coefficients in filter
    float *history     history array pointer

Returns float value giving the current output.

*************************************************************************/

float fir_filter(float input,float *coef,int n,float *history)
{
    int i;
    float *hist_ptr,*hist1_ptr,*coef_ptr;
    float output;

    hist_ptr = history;
    hist1_ptr = hist_ptr;              /* use for history update */
    coef_ptr = coef + n - 1;           /* point to last coef */

/* form output accumulation */
    output = *hist_ptr++ * (*coef_ptr--);
    for(i = 2 ; i < n ; i++) {
        *hist1_ptr++ = *hist_ptr;             /* update history array */
        output += (*hist_ptr++) * (*coef_ptr--);
    }
    output += input * (*coef_ptr);            /* input tap */
    *hist1_ptr = input;                       /* last history */

    return(output);
}
```

LISTING 4.1 Function **fir_filter(input,coef,n,history)**.

4.1.2 FIR Filter Coefficient Calculation

Because the stopband attenuation and passband ripple and the filter length are all specified as inputs to filter design programs, it is often difficult to determine the filter length required for a particular filter specification. Guessing the filter length will eventually reach a reasonable solution but can take a long time. For one stopband and one passband the following approximation for the filter length (N) of an optimal lowpass filter has been developed by Kaiser:

$$N = \frac{A_{stop} - 16}{29\Delta f} + 1 \tag{4.2}$$

where:

$$\Delta f = (f_{stop} - f_{pass}) / f_s$$

and A_{stop} is the minimum stopband attenuation (in dB) of the stopband from f_{stop} to $f_s/2$. The approximation for N is accurate within about 10 percent of the actual required filter length. For the Kaiser window design method, the passband error (δ_1) is equal to the stopband error (δ_2) and is related to the passband ripple (A_{max}) and stopband attenuation (in dB) as follows:

$$\delta_1 = 1 - 10^{-A_{max}/40}$$

$$\delta_2 = 10^{-A_{max}/20}$$

$$A_{max} = -40\log_{10}\left(1 - 10^{-A_{stop}/20}\right)$$

As a simple example, consider the following filter specifications, which specify a lowpass filter designed to remove the upper half of the signal spectrum:

Passband (f_{pass}):	$0\text{–}0.19 f_s$
Passband ripple (A_{max}):	< 0.2 dB
Stopband (f_{stop}):	$0.25 - 0.5 f_s$
Stopband Attenuation (A_{stop}):	> 40 dB

From these specifications

$$\delta_1 = 0.01145,$$

$$\delta_2 = 0.01,$$

$$\Delta f = 0.06.$$

The result of Equation (4.2) is $N = 37$. Greater stopband attenuation or a smaller transition band can be obtained with a longer filter. The filter coefficients are obtained by multiplying the Kaiser window coefficients by the ideal lowpass filter coefficients. The ideal lowpass coefficients for a very long odd length filter with a cutoff frequency of f_c are given by the following sinc function:

$$c_k = \frac{\sin(2 f_c k\pi)}{k\pi}. \tag{4.3}$$

Note that the center coefficient is $k = 0$ and the filter has even symmetry for all coefficients above $k = 0$. Very poor stopband attenuation would result if the above coefficients were truncated by using the 37 coefficients (effectively multiplying the *sinc function* by a rectangular window, which would have a stopband attenuation of about 13 dB). However, by multiplying these coefficients by the appropriate Kaiser window, the stopband and passband specifications can be realized. The symmetrical Kaiser window, w_k, is given by the following expression:

$$w_k = \frac{I_0\left\{\beta\sqrt{1-\left(1-\frac{2k}{N-1}\right)^2}\right\}}{I_0(\beta)}, \tag{4.4}$$

where $I_0(\beta)$ is a modified zero order Bessel function of the first kind, β is the Kaiser window parameter which determines the stopband attenuation. The empirical formula for β when A_{stop} is less than 50 dB is $\beta = 0.5842*(A_{stop} - 21)^{0.4} + 0.07886*(A_{stop} - 21)$. Thus, for a stopband attenuation of 40 dB, $\beta = 3.39532$. Listing 4.2 shows program KSRFIR.C, which can be used to calculate the coefficients of a FIR filter using the Kaiser window method. The length of the filter must be odd and bandpass; bandstop or highpass filters can also be designed. Figure 4.3(a) shows the frequency response of the resulting 37-point lowpass filter, and Figure 4.3(b) shows the frequency response of a 35-point lowpass filter designed using the Parks-McClellan program. The following computer dialog shows the results obtained using the REMEZ.C program:

```
   ...    REMEZ EXCHANGE FIR FILTER DESIGN PROGRAM ...

1: EXAMPLE1 -- LOWPASS FILTER
2: EXAMPLE2 -- BANDPASS FILTER
3: EXAMPLE3 -- DIFFERENTIATOR
4: EXAMPLE4 -- HILBERT TRANSFORMER
5: KEYBOARD -- GET INPUT PARAMETERS FROM KEYBOARD

selection [1 to 5] ? 5

number of coefficients [3 to 128] ? 35

Filter types are: 1=Bandpass, 2=Differentiator, 3=Hilbert

filter type [1 to 3] ? 1

number of bands [1 to 10] ? 2
Now inputting edge (corner) frequencies for 4 band edges

edge frequency for edge (corner) # 1 [0 to 0.5] ? 0

edge frequency for edge (corner) # 2 [0 to 0.5] ? .19

edge frequency for edge (corner) # 3 [0.19 to 0.5] ? .25
```

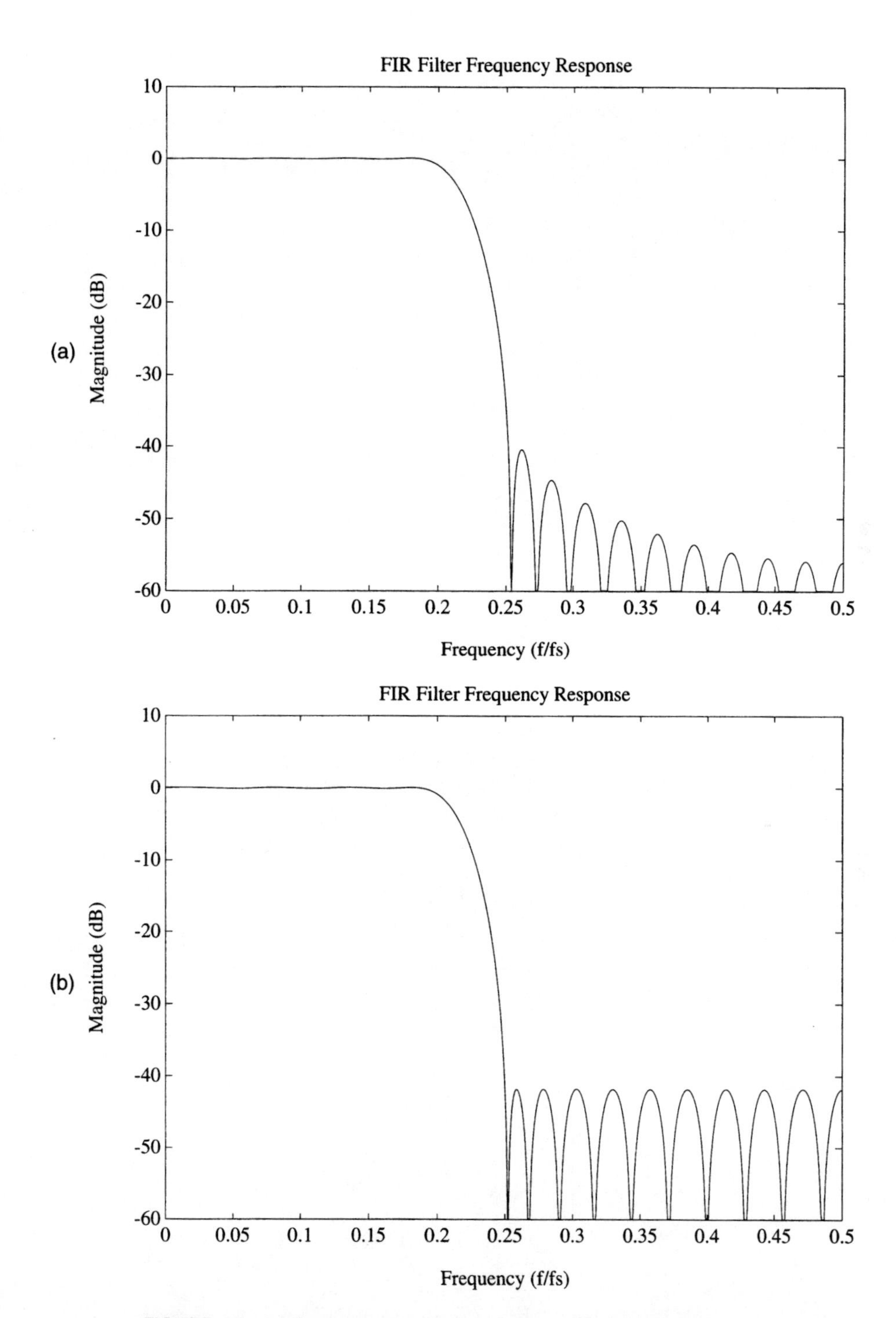

FIGURE 4.3 **(a)** Frequency response of 37 tap FIR filter designed using the Kaiser window method. **(b)** Frequency response of 35 tap FIR filter designed using the Parks-McClellan program.

```
edge frequency for edge (corner) # 4 [0.25 to 0.5] ? .5

gain of band # 1 [0 to 1000] ? 1

weight of band # 1 [0.1 to 100] ? 1

gain of band # 2 [0 to 1000] ? 0

weight of band # 2 [0.1 to 100] ? 1

#coeff = 35
Type = 1
#bands = 2
Grid = 16
E[1] = 0.00
E[2] = 0.19
E[3] = 0.25
E[4] = 0.50
Gain, wt[1] = 1.00 1.00
Gain, wt[2] = 0.00 1.00

Iteration 1 2 3 4 5 6 7
**********************************************************************
                        FINITE IMPULSE RESPONSE (FIR)
                        LINEAR PHASE DIGITAL FILTER DESIGN
                        REMEZ EXCHANGE ALGORITHM
                        BANDPASS FILTER
              FILTER LENGTH = 35
              ***** IMPULSE RESPONSE *****
                    H(  1) = -6.360096001e-003 = H(  35)
                    H(  2) = -7.662615827e-005 = H(  34)
                    H(  3) =  7.691285583e-003 = H(  33)
                    H(  4) =  5.056414595e-003 = H(  32)
                    H(  5) = -8.359812578e-003 = H(  31)
                    H(  6) = -1.040090568e-002 = H(  30)
                    H(  7) =  8.696002091e-003 = H(  29)
                    H(  8) =  2.017050147e-002 = H(  28)
                    H(  9) = -2.756078525e-003 = H(  27)
                    H( 10) = -3.003477728e-002 = H(  26)
                    H( 11) = -8.907503106e-003 = H(  25)
                    H( 12) =  4.171576865e-002 = H(  24)
                    H( 13) =  3.410815421e-002 = H(  23)
                    H( 14) = -5.073291821e-002 = H(  22)
                    H( 15) = -8.609754956e-002 = H(  21)
                    H( 16) =  5.791494030e-002 = H(  20)
                    H( 17) =  3.117008479e-001 = H(  19)
                    H( 18) =  4.402931165e-001 = H(  18)
```

```
                        BAND  1          BAND  2
LOWER BAND EDGE       0.00000000       0.25000000
UPPER BAND EDGE       0.19000000       0.50000000
DESIRED VALUE         1.00000000       0.00000000
WEIGHTING             1.00000000       1.00000000
DEVIATION             0.00808741       0.00808741
DEVIATION IN DB     -41.84380886     -41.84380886

EXTREMAL FREQUENCIES
  0.0156250     0.0520833     0.0815972     0.1093750     0.1371528
  0.1614583     0.1822917     0.1900000     0.2500000     0.2586806
  0.2777778     0.3038194     0.3298611     0.3576389     0.3854167
  0.4131944     0.4427083     0.4704861     0.5000000

FIR coefficients written to text file COEF.DAT
```

Note that the Parks-McClellan design achieved the specifications with two fewer coefficients, and the stopband attenuation is 1.8 dB better than the specification. Because the stopband attenuation, passband ripple, and filter length are all specified as inputs to the Parks-McClellan filter design program, it is often difficult to determine the filter length required for a particular filter specification. Guessing the filter length will eventually reach a reasonable solution but can take a long time. For one stopband and one passband, the following approximation for the filter length (N) of an optimal lowpass filter has been developed by Kaiser:

$$N = \frac{-20\log_{10}\sqrt{\delta_1\delta_2} - 13}{14.6\Delta f} + 1 \tag{4.5}$$

where:

$$\delta_1 = 1 - 10^{-A_{\max}/40}$$

$$\delta_2 = 10^{-A_{\text{stop}}/20}$$

$$\Delta f = (f_{\text{stop}} - f_{\text{pass}})/f_s$$

$A_{\max}$ is the total passband ripple (in dB) of the passband from 0 to f_{pass}. If the maximum of the magnitude response is 0 dB, then $A_{\max}$ is the maximum attenuation throughout the passband. A_{stop} is the minimum stopband attenuation (in dB) of the stopband from f_{stop} to $f_s/2$. The approximation for N is accurate within about 10 percent of the actual required filter length (usually on the low side). The ratio of the passband error (δ_1) to the stopband error (δ_2) is entered by choosing appropriate weights for each band. Higher weighting of stopbands will increase the minimum attenuation; higher weighting of the passband will decrease the passband ripple.

The coefficients for the Kaiser window design (variable name **fir_lpf37k**) and the Parks-McClellan design (variable name **fir_lpf35**) are contained in the include file FILTER.H.

```
/* Linear phase FIR filter coefficient computation using the Kaiser window
design method. Filter length is odd. */

#include <stdio.h>
#include <stdlib.h>
#include <string.h>
#include <math.h>
#include "rtdspc.h"

double get_float(char *title_string,double low_limit,double up_limit);
void filter_length(double att,double deltaf,int *nfilt,int *npair,double *beta);
double izero(double y);

void main()
{
    static float h[500], w[500], x[500];
    int eflag, filt_cat, npair, nfilt, n;
    double att, fa, fp, fa1, fa2, fp1, fp2, deltaf, d1, d2, fl, fu, beta;
    double fc, fm, pifc, tpifm, i, y, valizb;
    char ft_s[128];

    char  fp_s[] = "Passband edge frequency Fp";
    char  fa_s[] = "Stopband edge frequency Fa";
    char fp1_s[] = "Lower passband edge frequency Fp1";
    char fp2_s[] = "Upper passband edge frequency Fp2";
    char fa1_s[] = "Lower stopband edge frequency Fa1";
    char fa2_s[] = "Upper stopband edge frequency Fa2";

    printf("\nFilter type (lp, hp, bp, bs) ? ");
    gets(ft_s);
    strupr( ft_s );
    att = get_float("Desired stopband attenuation (dB)", 10, 200);
    filt_cat = 0;
    if( strcmp( ft_s, "LP" ) == 0 ) filt_cat = 1;
    if( strcmp( ft_s, "HP" ) == 0 ) filt_cat = 2;
    if( strcmp( ft_s, "BP" ) == 0 ) filt_cat = 3;
    if( strcmp( ft_s, "BS" ) == 0 ) filt_cat = 4;
    if(!filt_cat) exit(0);

    switch ( filt_cat ){
        case 1: case 2:
            switch ( filt_cat ){
                case 1:
                    fp = get_float( fp_s, 0, 0.5 );
                    fa = get_float( fa_s, fp, 0.5 ); break;
                case 2:
                    fa = get_float( fa_s, 0, 0.5 );
                    fp = get_float( fp_s, fa, 0.5 );
```

LISTING 4.2 Program KSRFIR to calculate FIR filter coefficients using the Kaiser window method. (*Continued*)

```
    }
    deltaf = (fa-fp) ; if(filt_cat == 2) deltaf = -deltaf;
    filter_length( att, deltaf, &nfilt, &npair, &beta );
    if( npair > 500 ){
        printf("\n*** Filter length %d is too large.\n", nfilt );
        exit(0);
    }
    printf("\n...filter length: %d ...beta: %f", nfilt, beta );
    fc = (fp + fa); h[npair] = fc;
    if ( filt_cat == 2 ) h[npair] = 1 - fc;
    pifc = PI * fc;
    for ( n=0; n < npair; n++){
        i = (npair - n);
        h[n] = sin(i * pifc) / (i * PI);
        if( filt_cat == 2 ) h[n] = - h[n];
    }
    break;
case 3: case 4:
    printf("\n—> Transition bands must be equal <—");
    do {
        eflag = 0;
        switch (filt_cat){
            case 3:
                fa1 = get_float( fa1_s,    0, 0.5);
                fp1 = get_float( fp1_s, fa1, 0.5);
                fp2 = get_float( fp2_s, fp1, 0.5);
                fa2 = get_float( fa2_s, fp2, 0.5); break;
            case 4:
                fp1 = get_float( fp1_s,    0, 0.5);
                fa1 = get_float( fa1_s, fp1, 0.5);
                fa2 = get_float( fa2_s, fa1, 0.5);
                fp2 = get_float( fp2_s, fa2, 0.5);
        }
        d1 = fp1 - fa1; d2 = fa2 - fp2;
        if ( fabs(d1 - d2) > 1E-5 ){
            printf( "\n...error...transition bands not equal\n");
            eflag = -1;
        }
    } while (eflag);
    deltaf = d1; if(filt_cat == 4) deltaf = -deltaf;
    filter_length( att, deltaf, &nfilt, &npair, &beta);
    if( npair > 500 ){
        printf("\n*** Filter length %d is too large.\n", nfilt );
        exit(0);
    }
    printf( "\n..filter length: %d ...beta: %f", nfilt, beta);
    fl = (fa1 + fp1) / 2; fu = (fa2 + fp2) / 2;
    fc = (fu - fl); fm = (fu + fl) / 2;
    h[npair] = 2 * fc; if( filt_cat == 4 ) h[npair] = 1 - 2 * fc;
```

LISTING 4.2 *(Continued)*

```
            pifc = PI * fc; tpifm = 2 * PI * fm;
            for (n = 0; n < npair; n++){
                i = (npair - n);
                h[n] = 2 * sin(i * pifc) * cos(i * tpifm) / (i * PI);
            if( filt_cat == 4) h[n] = -h[n];
            } break;
        default: printf( "\n## error\n" ); exit(0);
    }

/* Compute Kaiser window sample values */
    y = beta; valizb = izero(y);
    for (n = 0; n <= npair; n++){
        i = (n - npair);
        y = beta * sqrt(1 - (i / npair) * (i / npair));
        w[n] = izero(y) / valizb;
    }

/* first half of response */
    for(n = 0; n <= npair; n++) x[n] = w[n] * h[n];

    printf("\n--First half of coefficient set...remainder by symmetry--");
    printf("\n #     ideal        window      actual     ");
    printf("\n          coeff         value    filter coeff");
    for(n=0; n <= npair; n++){
        printf("\n %4d    %9.6f    %9.6f    %9.6f",n, h[n], w[n], x[n]);
    }
}

/* Use att to get beta (for Kaiser window function) and nfilt (always odd
    valued and = 2*npair +1) using Kaiser's empirical formulas */
void filter_length(double att,double deltaf,int *nfilt,int *npair,double *beta)
{
    *beta = 0;      /* value of beta if att < 21 */
    if(att >= 50) *beta = .1102 * (att - 8.71);
    if (att < 50 & att >= 21)
        *beta = .5842 * pow( (att-21), 0.4) + .07886 * (att - 21);
    *npair = (int)( (att - 8) / (29 * deltaf) );
    *nfilt = 2 * *npair +1;
}

/* Compute Bessel function Izero(y) using a series approximation */
double izero(double y){
    double s=1, ds=1, d=0;
    do{
        d = d + 2; ds = ds * (y*y)/(d*d);
        s = s + ds;
    } while( ds > 1E-7 * s);
    return(s);
}
```

LISTING 4.2 *(Continued)*

4.1.3 IIR Filter Function

Infinite impulse response (IIR) *filters* are realized by feeding back a weighted sum of past output values and adding this to a weighted sum of the previous and current input values. In terms of the structure shown in Figure 4.1, IIR filters have nonzero values for some or all of the a_n values. The major advantage of IIR filters compared to FIR filters is that a given order IIR filter can be made much more frequency selective than the same order FIR filter. In other words, IIR filters are computationally efficient. The disadvantage of the recursive realization is that IIR filters are much more difficult to design and implement. Stability, roundoff noise and sometimes phase nonlinearity must be considered carefully in all but the most trivial IIR filter designs.

The direct form IIR filter realization shown in Figure 4.1, though simple in appearance, can have severe response sensitivity problems because of coefficient quantization, especially as the order of the filter increases. To reduce these effects, the transfer function is usually decomposed into second order sections and then realized either as parallel or cascade sections (see chapter 1, section 1.3). In section 1.3.1 an IIR filter design and implementation method based on cascade decomposition of a transfer function into second order sections is described. The C language implementation shown in Listing 4.3 uses single-precision floating-point numbers in order to avoid coefficient quantization effects associated with fixed-point implementations that can cause instability and significant changes in the transfer function.

Figure 4.4 shows a block diagram of the cascaded second order IIR filter implemented by the **`iir_filter`** function shown in Listing 4.3. This realization is known as a direct form II realization because, unlike the structure shown in Figure 4.1, it has only

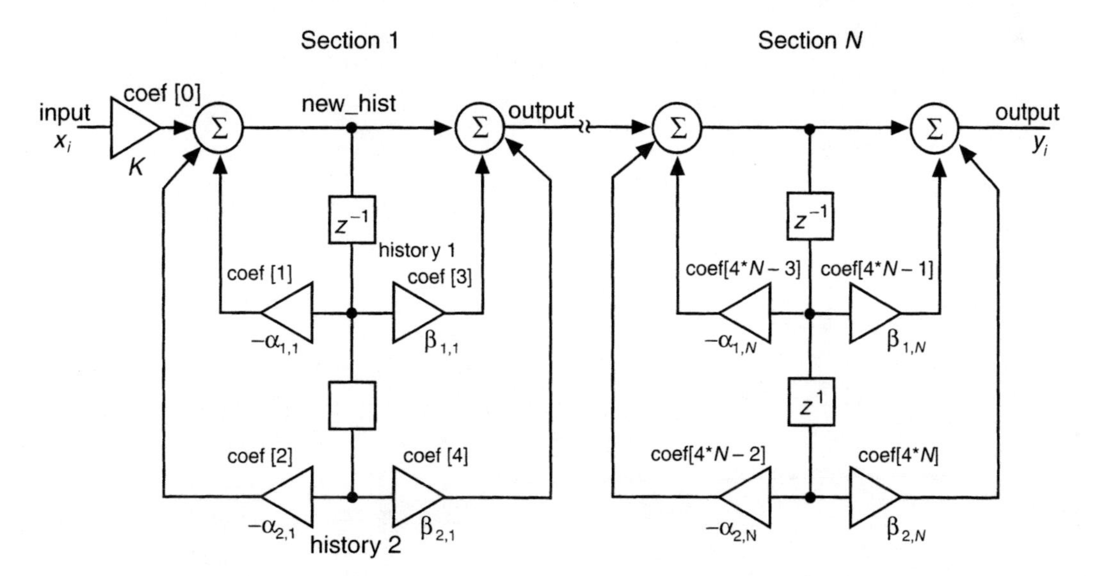

FIGURE 4.4 Block diagram of real-time IIR filter structure as implemented by function **`iir_filter`**.

```
/**************************************************************************

iir_filter - Perform IIR filtering sample by sample on floats

Implements cascaded direct form II second order sections.
Requires arrays for history and coefficients.
The length (n) of the filter specifies the number of sections.
The size of the history array is 2*n.
The size of the coefficient array is 4*n + 1 because
the first coefficient is the overall scale factor for the filter.
Returns one output sample for each input sample.

float iir_filter(float input,float *coef,int n,float *history)

    float input        new float input sample
    float *coef        pointer to filter coefficients
    int n              number of sections in filter
    float *history     history array pointer

Returns float value giving the current output.

*************************************************************************/

float iir_filter(float input,float *coef,int n,float *history)
{
    int i;
    float *hist1_ptr,*hist2_ptr,*coef_ptr;
    float output,new_hist,history1,history2;

    coef_ptr = coef;                  /* coefficient pointer */

    hist1_ptr = history;              /* first history */
    hist2_ptr = hist1_ptr + 1;              /* next history */

    output = input * (*coef_ptr++);      /* overall input scale factor */

    for(i = 0 ; i < n ; i++) {
        history1 = *hist1_ptr;               /* history values */
        history2 = *hist2_ptr;

        output = output - history1 * (*coef_ptr++);
        new_hist = output - history2 * (*coef_ptr++);     /* poles */

        output = new_hist + history1 * (*coef_ptr++);
        output = output + history2 * (*coef_ptr++);      /* zeros */

        *hist2_ptr++ = *hist1_ptr;
        *hist1_ptr++ = new_hist;
```

LISTING 4.3 Function **iir_filter(input,coef,n,history)**. (*Continued*)

```
        hist1_ptr++;
        hist2_ptr++;
    }

    return(output);
}
```

LISTING 4.3 *(Continued)*

two delay elements for each second-order section. This realization is canonic in the sense that the structure has the fewest adds (4), multiplies (4), and delay elements (2) for each second order section. This realization should be the most efficient for a wide variety of general purpose processors as well as many of the processors designed specifically for digital signal processing.

IIR filtering will be illustrated using a lowpass filter with similar specifications as used in the FIR filter design example in section 4.1.2. The only difference is that in the IIR filter specification, linear phase response is not required. Thus, the passband is 0 to $0.2\,f_s$ and the stopband is $0.25\,f_s$ to $0.5\,f_s$. The passband ripple must be less than 0.5 dB and the stopband attenuation must be greater than 40 dB. Because elliptic filters (also called *Cauer filters*) generally give the smallest transition bandwidth for a given order, an elliptic design will be used. After referring to the many elliptic filter tables, it is determined that a fifth order elliptic filter will meet the specifications. The elliptic filter tables in Zverev (1967) give an entry for a filter with a 0.28 dB passband ripple and 40.19 dB stopband attenuation as follows:

$\Omega_s = 1.3250$ (stopband start of normalized prototype)
$\sigma_0 = -0.5401$ (first order real pole)
$\sigma_1 = -0.5401$ (real part of first biquad section)
$\sigma_3 = -0.5401$ (real part of second biquad section)
$\Omega_1 = 1.0277$ (imaginary part of first biquad section)
$\Omega_2 = 1.9881$ (first zero on imaginary axis)
$\Omega_3 = 0.7617$ (imaginary part of second biquad section)
$\Omega_4 = 1.3693$ (second zero on imaginary axis)

As shown above, the tables in Zverev give the pole and zero locations (real and imaginary coordinates) of each biquad section. The two second-order sections each form a conjugate pole pair and the first-order section has a single pole on the real axis. Figure 4.5(a) shows the locations of the 5 poles and 4 zeros on the complex s-plane. By expanding the complex pole pairs, the s-domain transfer function of a fifth-order filter in terms of the above variables can be obtained. The z-domain coefficients are then determined using the bilinear transform (see Embree and Kimble, 1991). Figure 4.5(b) shows the locations of the poles and zeros on the complex z-plane. The resulting z-domain transfer function is as follows:

$$\frac{0.0553(1+z^{-1})}{1-0.436z^{-1}}\;\frac{1+0.704z^{-1}+z^{-2}}{1-0.523z^{-1}-0.86z^{-2}}\;\frac{1-0.0103z^{-1}+z^{-2}}{1-0.696z^{-1}-0.486z^{-2}}$$

Figure 4.6 shows the frequency response of this 5th order digital IIR filter.

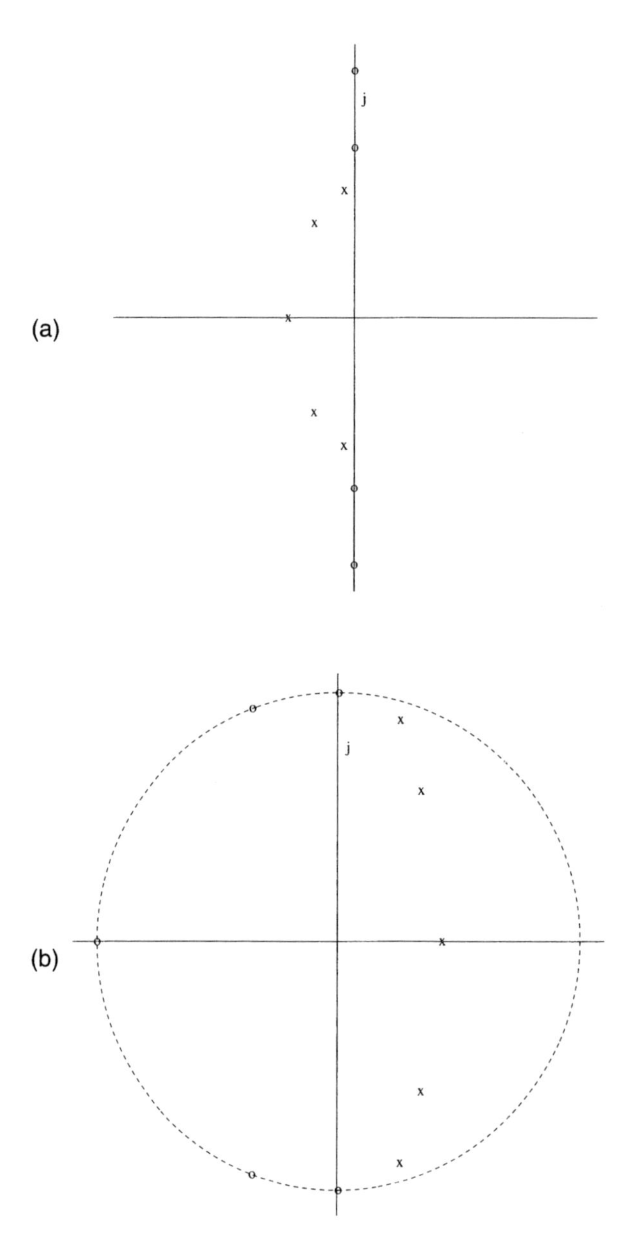

FIGURE 4.5 Pole-zero plot of fifth-order elliptic IIR lowpass filter. **(a)** *s*-plane representation of analog prototype fifth-order elliptic filter. Zeros are indicated by "o" and poles are indicated by "x". **(b)** *z*-plane representation of lowpass digital filter with cut-off frequency at 0.2 f_s. In each case, poles are indicated with "x" and zeros with "o".

The function **`iir_filter`** (shown in Listing 4.3) implements the direct form II cascade filter structure illustrated in Figure 4.4. Any number of cascaded second order sections can be implemented with one overall input (x_i) and one overall output (y_i). The coefficient array for the fifth order elliptic lowpass filter is as follows:

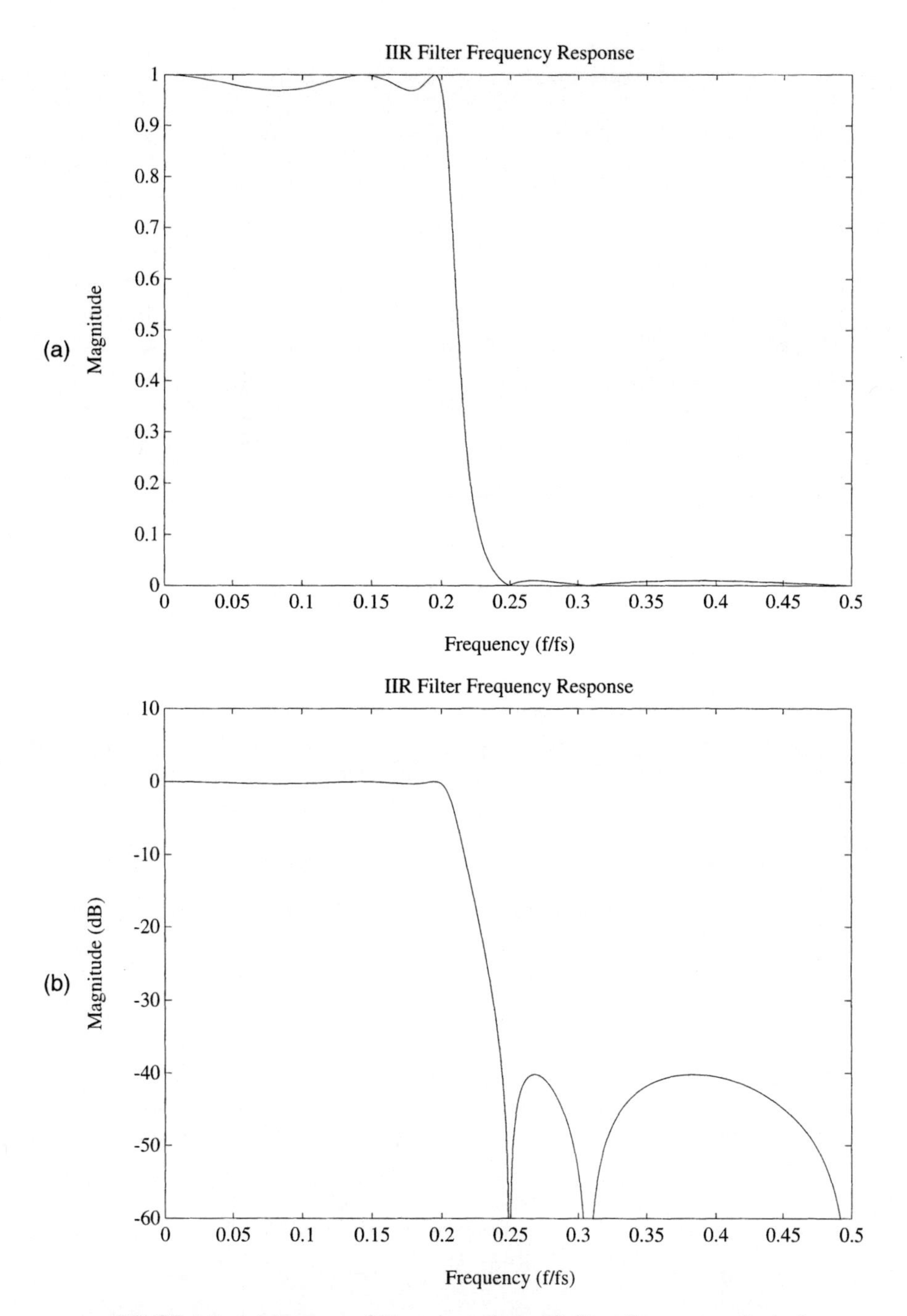

FIGURE 4.6 **(a)** Lowpass fifth-order elliptic IIR filter linear magnitude frequency response. **(b)** Lowpass fifth-order elliptic IIR filter frequency response. Log magnitude in decibels versus frequency.

```
float iir_lpf5[13] = {
     0.0552961603,
     -0,4363630712, 0.0000000000, 1.0000000000, 0.0000000000,
    -0.5233039260, 0.8604439497, 0.7039934993, 1.0000000000,
    -0.6965782046, 0.4860509932, -0.0103216320, 1.0000000000
};
```

The number of sections required for this filter is three, because the first-order section is implemented in the same way as the second-order sections, except that the second-order terms (the third and fifth coefficients) are zero. The coefficients shown above were obtained using the bilinear transform and are contained in the include file FILTER.H. The definition of this filter is, therefore, global to any module that includes FILTER.H. The **iir_filter** function filters the floating-point input sequence on a sample-by-sample basis so that one output sample is returned each time **iir_filter** is invoked. The history array is used to store the two history values required for each second-order section. The history data (two elements per section) is allocated by the calling function. The initial condition of the history variables is zero if **calloc** is used, because it sets all the allocated space to zero. If the history array is declared as static, most compilers initialize static space to zero. Other initial conditions can be loaded into the filter by allocating and initializing the history array before using the **iir_filter** function. The coefficients of the filter are stored with the overall gain constant (K) first, followed by the denominator coefficients that form the poles, and the numerator coefficients that form the zeros for each section. The input sample is first scaled by the K value, and then each second-order section is implemented. The four lines of code in the **iir_filter** function used to implement each second-order section are as follows:

```
output = output - history1 * (*coef_ptr++);
new_hist = output - history2 * (*coef_ptr++);    /* poles */

output = new_hist + history1 * (*coef_ptr++);
output = output + history2 * (*coef_ptr++);      /* zeros */
```

The **history1** and **history2** variables are the current history associated with the section and should be stored in floating-point registers (if available) for highest efficiency. The above code forms the new history value (the portion of the output which depends on the past outputs) in the variable **new_hist** to be stored in the history array for use by the next call to **iir_filter.** The history array values are then updated as follows:

```
*hist2_ptr++ = *hist1_ptr;
*hist1_ptr++ = new_hist;
hist1_ptr++;
hist2_ptr++;
```

This results in the oldest history value (***hist2_ptr**) being lost and updated with the more recent ***hist1_ptr** value. The **new_hist** value replaces the old

***hist1_ptr** value for use by the next call to **iir_filter.** Both history pointers are incremented twice to point to the next pair of history values to be used by the next second-order section.

4.1.4 Real-Time Filtering Example

Real-time filters are filters that are implemented so that a continuous stream of input samples can be filtered to generate a continuous stream of output samples. In many cases, real-time operation restricts the filter to operate on the input samples individually and generate one output sample for each input sample. Multiple memory accesses to previous input data are not possible, because only the current input is available to the filter at any given instant in time. Thus, some type of history must be stored and updated with each new input sample. The management of the filter history almost always takes a portion of the processing time, thereby reducing the maximum sampling rate which can be supported by a particular processor. The functions **fir_filter** and **iir_filter** are implemented in a form that can be used for real-time filtering. Suppose that the functions **getinput()** and **sendout()** return an input sample and generate an output sample at the appropriate time required by the external hardware. The following code can be used with the **iir_filter** function to perform continuous real-time filtering:

```
static float histi[6];
for(;;)
     sendout(iir_filter(getinput(),iir_lpf5,3,histi));
```

In the above infinite loop **for** statement, the total time required to execute the **in, iir_filter,** and **out** functions must be less than the filter sampling rate in order to insure that output and input samples are not lost. In a similar fashion, a continuous real-time FIR filter could be implemented as follows:

```
static float histf[34];
for(;;)
     sendout(fir_filter(getinput(),fir_lpf35,35,histf));
```

Source code for **sendout()** and **getinput()** interrupt driven input/output functions is available on the enclosed disk for several DSP processors. C code which emulates **getinput()** and **sendout()** real-time functions using disk read and write functions is also included on the disk and is shown in Listing 4.4. These routines can be used to debug real-time programs using a simpler and less expensive general purpose computer environment (IBM-PC or UNIX system, for example). The functions shown in Listing 4.4 read and write disk files containing floating-point numbers in an ASCII text format. The functions shown in Listings 4.5 and 4.6 read and write disk files containing fixed-point numbers in the popular WAV binary file format. The WAV file format is part of the Resource Interchange File Format (RIFF), which is popular on many multimedia platforms.

(text continues on page 158)

```
#include <stdlib.h>
#include <stdio.h>

/* getinput - get one sample from disk to simulate real-time input */

float getinput()
{
    static FILE *fp = NULL;
    float x;
/* open input file if not done in previous calls */
    if(!fp) {
        char s[80];
        printf("\nEnter input file name ? ");
        gets(s);
        fp = fopen(s,"r");
        if(!fp) {
            printf("\nError opening input file in GETINPUT\n");
            exit(1);
        }
    }
/* read data until end of file */
    if(fscanf(fp,"%f",&x) != 1) exit(1);
    return(x);
}

/* sendout - send sample to disk to simulate real-time output */

void sendout(float x)
{
    static FILE *fp = NULL;
/* open output file if not done in previous calls */
    if(!fp) {
        char s[80];
        printf("\nEnter output file name ? ");
        gets(s);
        fp = fopen(s,"w");
        if(!fp) {
            printf("\nError opening output file in SENDOUT\n");
            exit(1);
        }
    }
/* write the sample and check for errors */
    if(fprintf(fp,"%f\n",x) < 1) {
        printf("\nError writing output file in SENDOUT\n");
        exit(1);
    }
}
```

LISTING 4.4 Functions **sendout(output)** and **getinput()** used to emulate real-time input/output using ASCII text data files (contained in GETSEND.C).

```
#include <stdlib.h>
#include <stdio.h>
#include <string.h>
#include <math.h>
#include <conio.h>
#include "wavfmt.h"
#include "rtdspc.h"

/* code to get samples from a WAV type file format */

/* getinput - get one sample from disk to simulate realtime input */

/* input WAV format header with null init */
    WAVE_HDR win = { "", 0L };
    CHUNK_HDR cin = { "", 0L };
    DATA_HDR din = { "", 0L };
    WAVEFORMAT wavin = { 1, 1, 0L, 0L, 1, 8 };

/* global number of samples in data set */
    unsigned long int number_of_samples = 0;

float getinput()
{

    static FILE *fp_getwav = NULL;
    static channel_number = 0;
    short int int_data[4];          /* max of 4 channels can be read */
    unsigned char byte_data[4];     /* max of 4 channels can be read */
    short int j;
    int i;

/* open input file if not done in previous calls */
    if(!fp_getwav) {
        char s[80];
        printf("\nEnter input .WAV file name ? ");
        gets(s);
        fp_getwav = fopen(s,"rb");
        if(!fp_getwav) {
            printf("\nError opening *.WAV input file in GETINPUT\n");
            exit(1);
        }

/* read and display header information */
        fread(&win,sizeof(WAVE_HDR),1,fp_getwav);
        printf("\n%c%c%c%c",
            win.chunk_id[0],win.chunk_id[1],win.chunk_id[2],win.chunk_id[3]);
        printf("\nChunkSize = %ld bytes",win.chunk_size);
```

LISTING 4.5 Function **getinput()** used to emulate real-time input using WAV format binary data files (contained in GETWAV.C). (*Continued*)

```
        if(strnicmp(win.chunk_id,"RIFF",4) != 0) {
            printf("\nError in RIFF header\n");
            exit(1);
        }

        fread(&cin,sizeof(CHUNK_HDR),1,fp_getwav);
        printf("\n");
        for(i = 0 ; i < 8 ; i++) printf("%c",cin.form_type[i]);
        printf("\n");
        if(strnicmp(cin.form_type,"WAVEfmt ",8) != 0) {
            printf("\nError in WAVEfmt header\n");
            exit(1);
        }

        if(cin.hdr_size != sizeof(WAVEFORMAT)) {
            printf("\nError in WAVEfmt header\n");
            exit(1);
        }

        fread(&wavin,sizeof(WAVEFORMAT),1,fp_getwav);
        if(wavin.wFormatTag != WAVE_FORMAT_PCM) {
            printf("\nError in WAVEfmt header - not PCM\n");
            exit(1);
        }
        printf("\nNumber of channels = %d",wavin.nChannels);
        printf("\nSample rate = %ld",wavin.nSamplesPerSec);
        printf("\nBlock size of data = %d bytes",wavin.nBlockAlign);
        printf("\nBits per Sample = %d\n",wavin.wBitsPerSample);

/* check channel number and block size are good */
        if(wavin.nChannels > 4 || wavin.nBlockAlign > 8) {
            printf("\nError in WAVEfmt header - Channels/BlockSize\n");
            exit(1);
        }

        fread(&din,sizeof(DATA_HDR),1,fp_getwav);
        printf("\n%c%c%c%c",

din.data_type[0],din.data_type[1],din.data_type[2],din.data_type[3]);
        printf("\nData Size = %ld bytes",din.data_size);

/* set the number of samples (global) */
        number_of_samples = din.data_size/wavin.nBlockAlign;
        printf("\nNumber of Samples per Channel = %ld\n",number_of_samples);

        if(wavin.nChannels > 1) {
          do {
            printf("\nError Channel Number [0..%d] - ",wavin.nChannels-1);
```

LISTING 4.5 *(Continued)*

```
            i = getche() - '0';
            if(i < (4-'0')) exit(1);
         } while(i < 0 || i >= wavin.nChannels);
         channel_number = i;
        }
    }

/* read data until end of file */
    if(wavin.wBitsPerSample == 16) {
        if(fread(int_data,wavin.nBlockAlign,1,fp_getwav) != 1) {
            flush(); /* flush the output when input runs out */
            exit(1);
        }
        j = int_data[channel_number];
    }
    else {
        if(fread(byte_data,wavin.nBlockAlign,1,fp_getwav) != 1) {
            flush(); /* flush the output when input runs out */
            exit(1);
        }
        j = byte_data[channel_number];
        j ^= 0x80;
        j <<= 8;
    }

    return((float)j);
}
```

LISTING 4.5 (*Continued*)

```
#include <stdlib.h>
#include <stdio.h>
#include <string.h>
#include <math.h>
#include "wavfmt.h"
#include "rtdspc.h"

/* code to send samples to a WAV type file format */

/* define BITS16 if want to use 16 bit samples */

/* sendout - send sample to disk to simulate realtime output */

    static FILE *fp_sendwav = NULL;
    static DWORD samples_sent = 0L; /* used by flush for header */
```

LISTING 4.6 Functions `sendout(output)` and `flush()` used to emulate real-time output using WAV format binary data files (contained in SENDWAV.C). (*Continued*)

```
/* WAV format header init */
    static WAVE_HDR wout = { "RIFF", 0L }; /* fill size at flush */
    static CHUNK_HDR cout = { "WAVEfmt " , sizeof(WAVEFORMAT) };
    static DATA_HDR dout = { "data" , 0L }; /* fill size at flush */
    static WAVEFORMAT wavout = { 1, 1, 0L, 0L, 1, 8 };

    extern WAVE_HDR win;
    extern CHUNK_HDR cin;
    extern DATA_HDR din;
    extern WAVEFORMAT wavin;

void sendout(float x)
{
    int BytesPerSample;
    short int j;

/* open output file if not done in previous calls */
    if(!fp_sendwav) {
        char s[80];
        printf("\nEnter output *.WAV file name ? ");
        gets(s);
        fp_sendwav = fopen(s,"wb");
        if(!fp_sendwav) {
            printf("\nError opening output *.WAV file in SENDOUT\n");
            exit(1);
        }
/* write out the *.WAV file format header */

#ifdef BITS16
        wavout.wBitsPerSample = 16;
        wavout.nBlockAlign = 2;
        printf("\nUsing 16 Bit Samples\n");
#else
        wavout.wBitsPerSample = 8;
#endif

        wavout.nSamplesPerSec = SAMPLE_RATE;
        BytesPerSample = (int)ceil(wavout.wBitsPerSample/8.0);
        wavout.nAvgBytesPerSec = BytesPerSample*wavout.nSamplesPerSec;

        fwrite(&wout,sizeof(WAVE_HDR),1,fp_sendwav);
        fwrite(&cout,sizeof(CHUNK_HDR),1,fp_sendwav);
        fwrite(&wavout,sizeof(WAVEFORMAT),1,fp_sendwav);
        fwrite(&dout,sizeof(DATA_HDR),1,fp_sendwav);
    }
/* write the sample and check for errors */

/* clip output to 16 bits */
    j = (short int)x;
```

LISTING 4.6 *(Continued)*

```
    if(x > 32767.0) j = 32767;
    else if(x < -32768.0) j = -32768;

#ifdef BITS16
    j ^= 0x8000;

    if(fwrite(&j,sizeof(short int),1,fp_sendwav) != 1) {
        printf("\nError writing 16 Bit output *.WAV file in SENDOUT\n");
        exit(1);
    }
#else
/* clip output to 8 bits */
    j = j >> 8;
    j ^= 0x80;

    if(fputc(j,fp_sendwav) == EOF) {
        printf("\nError writing output *.WAV file in SENDOUT\n");
        exit(1);
    }
#endif

    samples_sent++;
}

/* routine for flush - must call this to update the WAV header */

void flush()
{
    int BytesPerSample;

    BytesPerSample = (int)ceil(wavout.wBitsPerSample/8.0);
    dout.data_size=BytesPerSample*samples_sent;

    wout.chunk_size=
        dout.data_size+sizeof(DATA_HDR)+sizeof(CHUNK_HDR)+sizeof(WAVEFORMAT);

/* check for an input WAV header and use the sampling rate, if valid */
    if(strnicmp(win.chunk_id,"RIFF",4) == 0 && wavin.nSamplesPerSec != 0) {
        wavout.nSamplesPerSec = wavin.nSamplesPerSec;
        wavout.nAvgBytesPerSec = BytesPerSample*wavout.nSamplesPerSec;
    }

    fseek(fp_sendwav,0L,SEEK_SET);
    fwrite(&wout,sizeof(WAVE_HDR),1,fp_sendwav);
    fwrite(&cout,sizeof(CHUNK_HDR),1,fp_sendwav);
    fwrite(&wavout,sizeof(WAVEFORMAT),1,fp_sendwav);
    fwrite(&dout,sizeof(DATA_HDR),1,fp_sendwav);
}
```

LISTING 4.6 *(Continued)*

4.2 FILTERING TO REMOVE NOISE

Noise is generally unwanted and can usually be reduced by some type of filtering. Noise can be highly correlated with the signal or in a completely different frequency band, in which case it is uncorrelated. Some types of noise are impulsive in nature and occur relatively infrequently, while other types of noise appear as narrowband tones near the signal of interest. The most common type of noise is wideband thermal noise, which originates in the sensor or the amplifying electronic circuits. Such noise can often be considered white Gaussian noise, implying that the power spectrum is flat and the distribution is normal. The most important considerations in deciding what type of filter to use to remove noise are the type and characteristics of the noise. In many cases, very little is known about the noise process contaminating the digital signal and it is usually costly (in terms of time and/or money) to find out more about it. One method to study the noise performance of a digital system is to generate a model of the signal and noise and simulate the system performance in this ideal condition. System noise simulation is illustrated in the next two sections. The simulated performance can then be compared to the system performance with real data or to a theoretical model.

4.2.1 Gaussian Noise Generation

The function **gaussian** (shown in Listing 4.7) is used for noise generation and is contained in the FILTER.C source file. The function has no arguments and returns a single random floating-point number. The standard C library function **rand** is called to generate uniformly distributed numbers. The function **rand** normally returns integers from 0 to some maximum value (a defined constant, **RAND_MAX**, in ANSI implementations). As shown in Listing 4.7, the integer values returned by **rand** are converted to **float** values to be used by **gaussian**. Although the random number generator provided with most C compilers gives good random numbers with uniform distributions and long periods, if the random number generator is used in an application that requires truly random, uncorrelated sequences, the generator should be checked carefully. If the **rand** function is in question, a standard random number generator can be easily written in C (see Park and Miller, 1988). The function **gaussian** returns a zero mean random number with a unit variance and a Gaussian (or normal) distribution. It uses the Box-Muller method (see Knuth, 1981; or Press, Flannary, Teukolsky, and Vetterling, 1987) to map a pair of independent uniformly distributed random variables to a pair of Gaussian random variables. The function **rand** is used to generate the two uniform variables **v1** and **v2** from –1 to +1, which are transformed using the following statements:

```
r = v1*v1 + v2*v2;
fac = sqrt(-2.*log(r)/r);
gstore = v1*fac;
gaus = v2*fac;
```

The **r** variable is the radius squared of the random point on the (**v1, v2**) plane. In the **gaussian** function, the **r** value is tested to insure that it is always less than 1 (which it

```
/*************************************************************************

gaussian - generates zero mean unit variance Gaussian random numbers

Returns one zero mean unit variance Gaussian random numbers as a double.
Uses the Box-Muller transformation of two uniform random numbers to
Gaussian random numbers.

*************************************************************************/

float gaussian()
{
    static int ready = 0;          /* flag to indicated stored value */
    static float gstore;           /* place to store other value */
    static float rconst1 = (float)(2.0/RAND_MAX);
    static float rconst2 = (float)(RAND_MAX/2.0);
    float v1,v2,r,fac,gaus;

/* make two numbers if none stored */
    if(ready == 0) {
        do {
            v1 = (float)rand() - rconst2;
            v2 = (float)rand() - rconst2;
            v1 *= rconst1;
            v2 *= rconst1;
            r = v1*v1 + v2*v2;
        } while(r > 1.0f);           /* make radius less than 1 */

/* remap v1 and v2 to two Gaussian numbers */
        fac = sqrt(-2.0f*log(r)/r);
        gstore = v1*fac;           /* store one */
        gaus = v2*fac;             /* return one */
        ready = 1;                 /* set ready flag */
    }

    else {
        ready = 0;      /* reset ready flag for next pair */
        gaus = gstore; /* return the stored one */
    }
    return(gaus);
}
```

LISTING 4.7 Function **gaussian()**.

usually is), so that the region uniformly covered by (**v1, v2**) is a circle and so that **log(r)** is always negative and the argument for the square root is positive. The variables **gstore** and **gaus** are the resulting independent Gaussian random variables. Because **gaussian** must return one value at a time, the **gstore** variable is a **static** floating-point variable used to store the **v1*fac** result until the next call to **gaussian**.

The **static** integer variable **ready** is used as a flag to indicate if **gstore** has just been stored or if two new Gaussian random numbers should be generated.

4.2.2 Signal-to-Noise Ratio Improvement

One common application of digital filtering is *signal-to-noise ratio enhancement.* If the signal has a limited bandwidth and the noise has a spectrum that is broad, then a filter can be used to remove the part of the noise spectrum that does not overlap the signal spectrum. If the filter is designed to match the signal perfectly, so that the maximum amount of noise is removed, then the filter is called a *matched* or *Wiener filter.* Wiener filtering is briefly discussed in section 1.7.1 of chapter 1.

Figure 4.7 shows a simple example of filtering a single tone with added white noise. The MKGWN program (see Listing 4.8) was used to add Gaussian white noise with a standard deviation of 0.2 to a sine wave at a 0.05 f_s frequency as shown in Figure 4.7(a). The standard deviation of the sine wave signal alone can be easily found to be 0.7107. Because the standard deviation of the added noise is 0.2, the signal-to-noise ratio of the noisy signal is 3.55 or 11.0 dB. Figure 4.7(b) shows the result of applying the 35-tap lowpass FIR filter to the noisy signal. Note that much of the noise is still present but is smaller and has predominantly low frequency components. By lowpass filtering the 250 noise samples added to the sine wave separately, the signal-to-noise ratio of Figure 4.7(b) can be estimated to be 15 dB. Thus, the filtering operation improved the signal-to-noise ratio by 4 dB.

4.3 SAMPLE RATE CONVERSION

Many signal processing applications require that the output sampling rate be different than the input sampling rate. Sometimes one section of a system can be made more efficient if the sampling rate is lower (such as when simple FIR filters are involved or in data transmission). In other cases, the sampling rate must be increased so that the spectral details of the signal can be easily identified. In either case, the input sampled signal must be resampled to generate a new output sequence with the same spectral characteristics but at a different sampling rate. Increasing the sampling rate is called *interpolation* or *upsampling.* Reducing the sampling rate is called *decimation* or *downsampling.* Normally, the sampling rate of a band limited signal can be interpolated or decimated by integer ratios such that the spectral content of the signal is unchanged. By cascading interpolation and decimation, the sampling rate of a signal can be changed by any rational fraction, P/M, where P is the integer interpolation ratio and M is the integer decimation ratio. Interpolation and decimation can be performed using filtering techniques (as described in this section) or by using the fast Fourier transform (see section 4.4.2).

Decimation is perhaps the simplest resampling technique because it involves reducing the number of samples per second required to represent a signal. If the input signal is strictly band-limited such that the signal spectrum is zero for all frequencies above $f_s/(2M)$, then decimation can be performed by simply retaining every Mth sample and

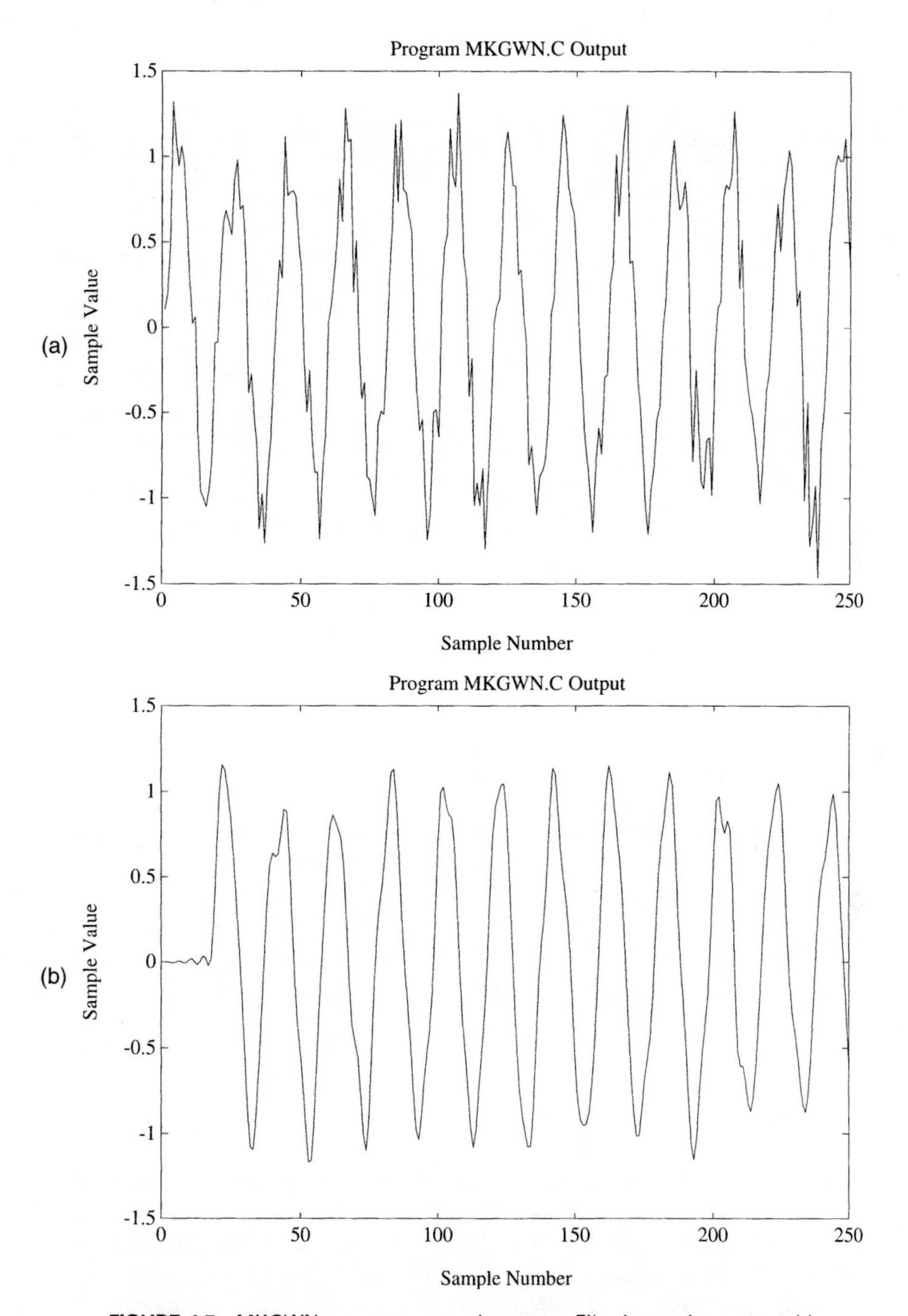

FIGURE 4.7 MKGWN program example output. Filtering a sine wave with added noise (frequency = 0.05). **(a)** Unfiltered version with Gaussian noise (standard deviation = 0.2). **(b)** Output after lowpass filtering with 35-point FIR filter.

```
#include <stdlib.h>
#include <stdio.h>
#include <string.h>
#include <math.h>
#include "rtdspc.h"
#include "filter.h"

/**************************************************************************

MKGWN.C - Gaussian Noise Filter Example

This program performs filters a sine wave with added Gaussian noise
It performs the filtering to implement a 35 point FIR filter
(stored in variable fir_lpf35) on an generated signal.
The filter is a LPF with 40 dB out of band rejection. The 3 dB
point is at a relative frequency of approximately .25*fs.

**************************************************************************/

float sigma = 0.2;

void main()
{
 int         i, j;
 float x;
 static float hist[34];
 for(i = 0 ; i < 250 ; i++) {
     x = sin(0.05*2*PI*i) + sigma*gaussian();
     sendout(fir_filter(x,fir_lpf35,35,hist));
 }
}
```

LISTING 4.8 Program MKGWN to add Gaussian white noise to cosine wave and then perform FIR filtering.

discarding the $M - 1$ samples in between. Unfortunately, the spectral content of a signal above $f_s/(2M)$ is rarely zero, and the aliasing caused by the simple decimation almost always causes trouble. Even when the desired signal is zero above $f_s/(2M)$, some amount of noise is usually present that will alias into the lower frequency signal spectrum. Aliasing due to decimation can be avoided by lowpass filtering the signal before the samples are decimated. For example, when $M = 2$, the 35-point lowpass FIR filter introduced in section 4.1.2 can be used to eliminate almost all spectral content above $0.25f_s$ (the attenuation above $0.25\, f_s$ is greater than 40 dB). A simple decimation program could then be used to reduce the sampling rate by a factor of two. An IIR lowpass filter (discussed in section 4.1.3) could also be used to eliminate the frequencies above $f_s/(2M)$ as long as linear phase response is not required.

4.3.1 FIR Interpolation

Interpolation is the process of computing new samples in the intervals between existing data points. Classical interpolation (used before calculators and computers) involves estimating the value of a function between existing data points by fitting the data to a low-order polynomial. For example, *linear* (first-order) or *quadratic* (second-order) polynomial interpolation is often used. The primary attraction of polynomial interpolation is computational simplicity. The primary disadvantage is that in signal processing, the input signal must be restricted to a very narrow band so that the output will not have a large amount of aliasing. Thus, band-limited interpolation using digital filters is usually the method of choice in digital signal processing. Band-limited interpolation by a factor P:1 (see Figure 4.8 for an illustration of 3:1 interpolation) involves the following conceptual steps:

(1) Make an output sequence P times longer than the input sequence. Place the input sequence in the output sequence every P samples and place $P - 1$ zero values between each input sample. This is called *zero-packing* (as opposed to *zero-padding*). The zero values are located where the new interpolated values will appear. The effect of zero-packing on the input signal spectrum is to replicate the spectrum P times within the output spectrum. This is illustrated in Figure 4.8(a) where the output sampling rate is three times the input sampling rate.

(2) Design a lowpass filter capable of attenuating the undesired $P - 1$ spectra above the original input spectrum. Ideally, the passband should be from 0 to $f_s'/(2P)$ and the stopband should be from $f_s'/(2P)$ to $f_s'/2$ (where f_s' is the filter sampling rate that is P times the input sampling rate). A more practical interpolation filter has a transition band centered about $f_s'/(2P)$. This is illustrated in Figure 4.8(b). The passband gain of this filter must be equal to P to compensate for the inserted zeros so that the original signal amplitude is preserved.

(3) Filter the zero-packed input sequence using the interpolation filter to generate the final P:1 interpolated signal. Figure 4.8(c) shows the resulting 3:1 interpolated spectrum. Note that the two repeated spectra are attenuated by the stopband attenuation of the interpolation filter. In general, the stopband attenuation of the filter must be greater than the signal-to-noise ratio of the input signal in order for the interpolated signal to be a valid representation of the input.

4.3.2 Real-Time Interpolation Followed by Decimation

Figure 4.8(d) illustrates 2:1 decimation after the 3:1 interpolation, and shows the spectrum of the final signal, which has a sampling rate 1.5 times the input sampling rate. Because no lowpass filtering (other than the filtering by the 3:1 interpolation filter) is performed before the decimation shown, the output signal near $f_s''/2$ has an unusually shaped power spectrum due to the aliasing of the 3:1 interpolated spectrum. If this aliasing causes a problem in the system that processes the interpolated output signal, it can be

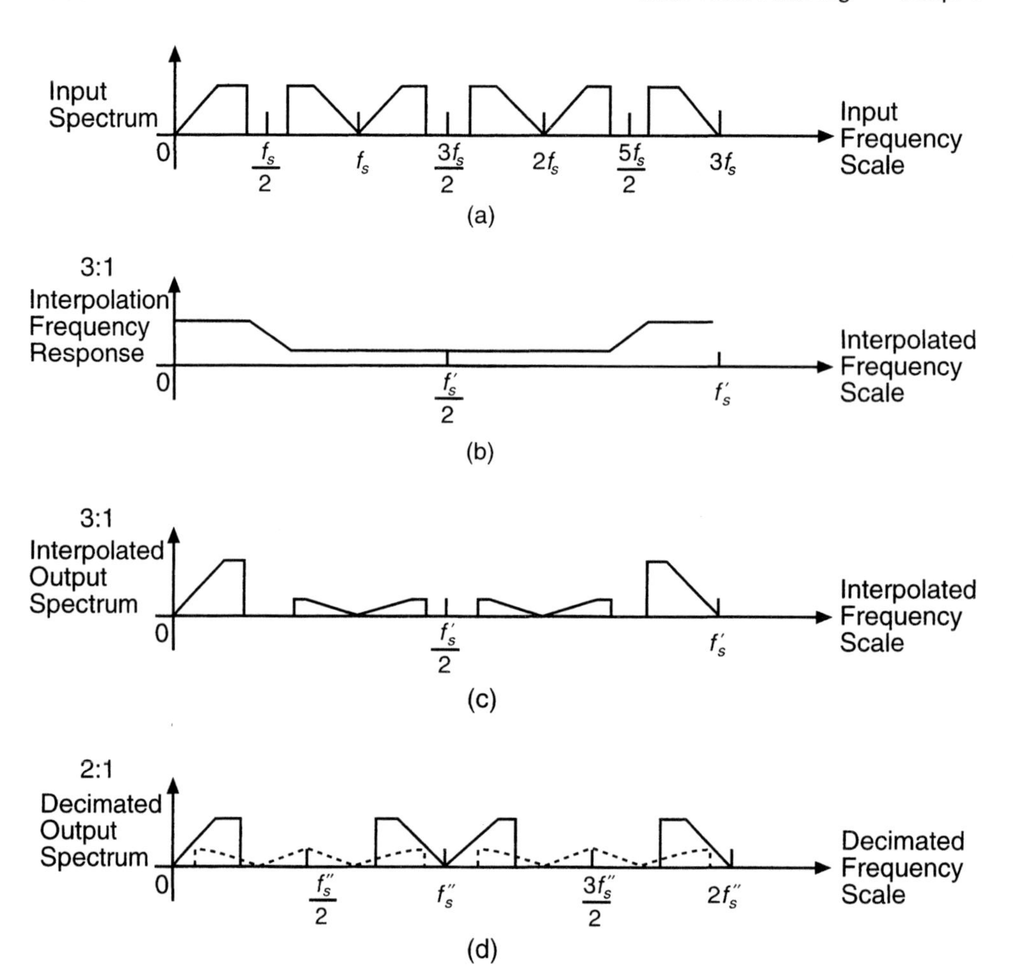

FIGURE 4.8 Illustration of 3:1 interpolation followed by 2:1 decimation. The aliased input spectrum in the decimated output is shown with a dashed line. **(a)** Example real input spectrum. **(b)** 3:1 interpolation filter response ($f_s' = 3f_s$). **(c)** 3:1 interpolated spectrum. **(d)** 2:1 decimated output ($f_s'' = f_s'/2$).

eliminated by either lowpass filtering the signal before decimation or by designing the interpolation filter to further attenuate the replicated spectra.

The interpolation filter used to create the interpolated values can be an IIR or FIR lowpass filter. However, if an IIR filter is used the input samples are not preserved exactly because of the nonlinear phase response of the IIR filter. FIR interpolation filters can be designed such that the input samples are preserved, which also results in some computational savings in the implementation. For this reason, only the implementation of FIR interpolation will be considered further. The FIR lowpass filter required for interpolation can be designed using the simpler windowing techniques. In this section, a Kaiser window is used to design 2:1 and 3:1 interpolators. The FIR filter length must be odd so that the filter delay is an integer number of samples and the input samples can be preserved. The passband and stopband must be specified such that the center coefficient of the filter is unity (the filter gain will be P) and P coefficients on each side of the filter center are zero. This insures that the original input samples are preserved, because the result of all the multiplies in the convolution is zero, except for the center filter coefficient that gives the input sample. The other $P - 1$ output samples between each original input sample are created by convolutions with the other coefficients of the filter. The following passband and stopband specifications will be used to illustrate a P:1 interpolation filter:

Passband frequencies:	$0\text{–}0.8 f_s/(2P)$
Stopband frequencies:	$1.2 f_s/(2P)\text{–}0.5 f_s$
Passband gain:	P
Passband ripple:	< 0.03 dB
Stopband attenuation:	> 56 dB

The filter length was determined to be $16P - 1$ using Equation (4.2) (rounding to the nearest odd length) and the passband and stopband specifications. Greater stopband attenuation or a smaller transition band can be obtained with a longer filter. The interpolation filter coefficients are obtained by multiplying the Kaiser window coefficients by the ideal lowpass filter coefficients. The ideal lowpass coefficients for a very long odd length filter with a cutoff frequency of $f_s/2P$ are given by the following sinc function:

$$c_k = \frac{P \sin(k\pi / P)}{k\pi}. \tag{4.6}$$

Note that the original input samples are preserved, because the coefficients are zero for all $k = nP$, where n is an integer greater than zero and $c_0 = 1$. Very poor stopband attenuation would result if the above coefficients were truncated by using the $16P - 1$ coefficients where $|k| < 8P$. However, by multiplying these coefficients by the appropriate Kaiser window, the stopband and passband specifications can be realized. The symmetrical Kaiser window, w_k, is given by the following expression:

$$w_k = \frac{I_0\left\{\beta\sqrt{1-\left(\frac{2k}{N-1}\right)^2}\right\}}{I_0(\beta)}, \tag{4.7}$$

where $I_0(\beta)$ is a modified zero order Bessel function of the first kind, β is the Kaiser window parameter which determines the stopband attenuation and N in equation (4.7) is $16P + 1$. The empirical formula for β when A_{stop} is greater than 50 dB is $\beta = 0.1102*(A_{stop} - 8.71)$. Thus, for a stopband attenuation of 56 dB, $\beta = 5.21136$. Figure 4.9(a) shows the frequency response of the resulting 31-point 2:1 interpolation filter, and Figure 4.9(b) shows the frequency response of the 47-point 3:1 interpolation filter.

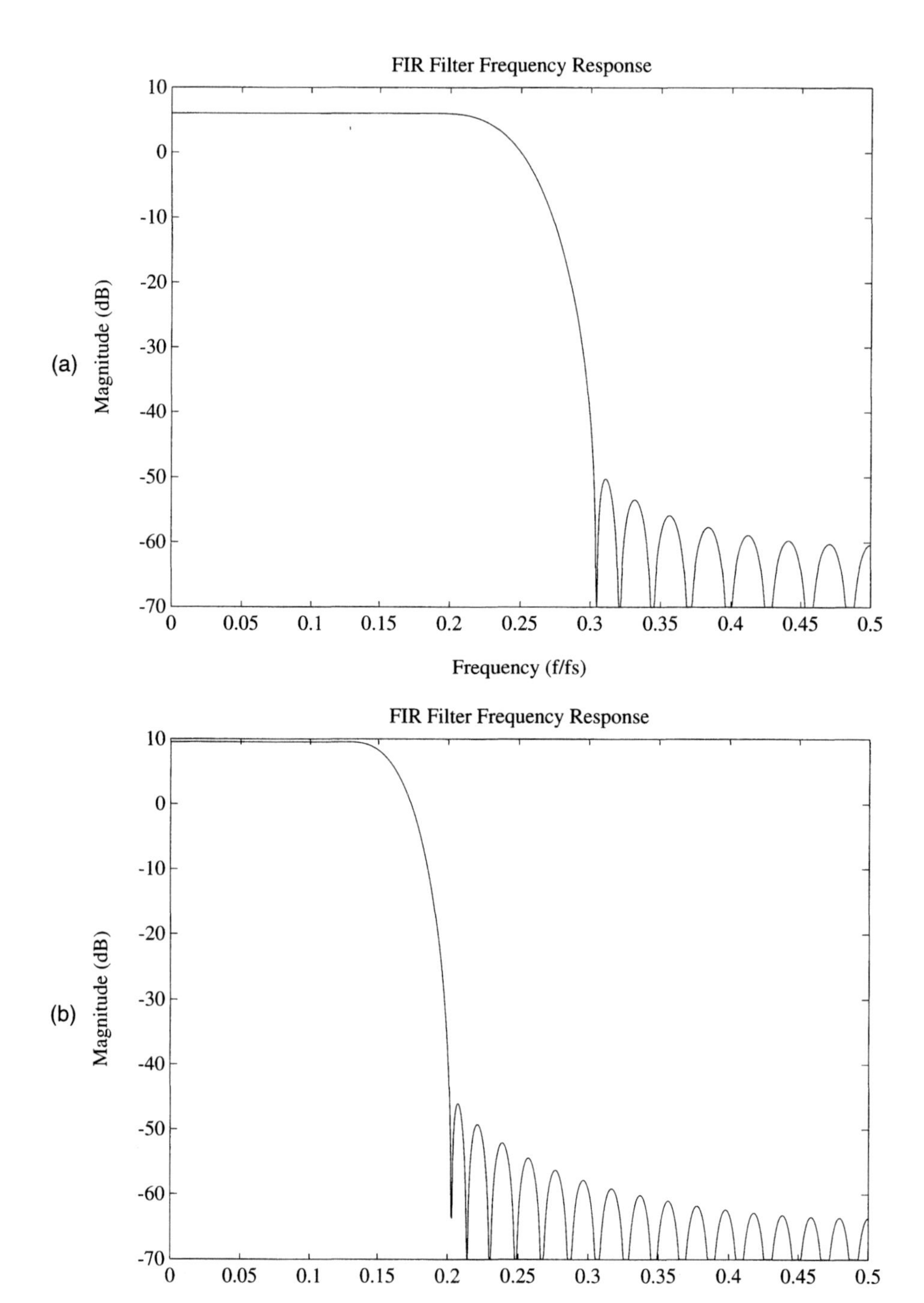

FIGURE 4.9 **(a)** Frequency response of 31-point FIR 2:1 interpolation filter (gain = 2 or 6 dB). **(b)** Frequency response of 47-point FIR 3:1 interpolation filter (gain = 3 or 9.54 dB).

4.3.3 Real-Time Sample Rate Conversion

Listing 4.9 shows the example interpolation program INTERP3.C, which can be used to interpolate a signal by a factor of 3. Two coefficient arrays are initialized to have the decimated coefficients each with 16 coefficients. Each of the coefficient sets are then used individually with the **fir_filter** function to create the interpolated values to be sent to **sendout()**. The original input signal is copied without filtering to the output every *P* sample (where *P* is 3). Thus, compared to direct filtering using the 47-point original filter, 15 multiplies for each input sample are saved when interpolation is performed using INTERP3. Note that the rate of output must be exactly three times the rate of input for this program to work in a real-time system.

```
#include <stdlib.h>
#include <stdio.h>
#include <string.h>
#include <math.h>
#include "rtdspc.h"

/**************************************************************************

INTERP3.C - PROGRAM TO DEMONSTRATE 3:1 FIR FILTER INTERPOLATION
            USES TWO INTERPOLATION FILTERS AND MULTIPLE CALLS TO THE
            REAL TIME FILTER FUNCTION fir_filter().

**************************************************************************/

main()
{
    int i;
    float signal_in;
/* interpolation coefficients for the decimated filters */
    static float coef31[16],coef32[16];
/* history arrays for the decimated filters */
    static float hist31[15],hist32[15];

/* 3:1 interpolation coefficients, PB 0-0.133, SB 0.2-0.5 */
    static float interp3[47] = {
   -0.00178662, -0.00275941, 0.,  0.00556927,  0.00749929, 0.,
   -0.01268113, -0.01606336, 0.,  0.02482278,  0.03041984, 0.,
   -0.04484686, -0.05417098, 0.,  0.07917613,  0.09644332, 0.,
   -0.14927754, -0.19365910, 0.,  0.40682136,  0.82363913, 1.0,
    0.82363913,  0.40682136, 0., -0.19365910, -0.14927754, 0.,
    0.09644332,  0.07917613, 0., -0.05417098, -0.04484686, 0.,
    0.03041984,  0.02482278, 0., -0.01606336, -0.01268113, 0.,
```

LISTING 4.9 Example INTERP3.C program for 3:1 FIR interpolation. (*Continued*)

```
    0.00749928, 0.00556927, 0., -0.00275941, -0.00178662
            };

    for(i = 0 ; i < 16 ; i++) coef31[i] = interp3[3*i];

    for(i = 0 ; i < 16 ; i++) coef32[i] = interp3[3*i+1];

/* make three samples for each input */
    for(;;) {
        signal_in = getinput();
        sendout(hist31[7]); /* delayed input */
        sendout(fir_filter(signal_in,coef31,16,hist31));
        sendout(fir_filter(signal_in,coef32,16,hist32));
    }
}
```

LISTING 4.9 *(Continued)*

Figure 4.10 shows the result of running the INTERP3.C program on the WAVE3.DAT data file contained on the disk (the sum of frequencies 0.01, 0.02 and 0.4). Figure 4.10(a) shows the original data. The result of the 3:1 interpolation ratio is shown in Figure 4.10(b). Note that the definition of the highest frequency in the original data set ($0.4\,f_s$) is much improved, because in Figure 4.10(b) there are 7.5 samples per cycle of the highest frequency. The startup effects and the 23 sample delay of the 47-point interpolation filter is also easy to see in Figure 4.10(b) when compared to Figure 4.10(a).

4.4 FAST FILTERING ALGORITHMS

The FFT is an extremely useful tool for spectral analysis. However, another important application for which FFTs are often used is fast convolution. The formulas for convolution were given in chapter 1. Most often a relatively short sequence 20 to 200 points in length (for example, an FIR filter) must be convolved with a number of longer input sequences. The input sequence length might be 1,000 samples or greater and may be changing with time as new data samples are taken.

One method for computation given this problem is straight implementation of the time domain convolution equation as discussed extensively in chapter 4. The number of real multiplies required is $M * (N - M + 1)$, where N is the input signal size and M is the length of the FIR filter to be convolved with the input signal. There is an alternative to this rather lengthy computation method—the convolution theorem. The convolution theorem states that time domain convolution is equivalent to multiplication in the frequency domain. The convolution equation above can be rewritten in the frequency domain as follows:

$$Y(k) = H(k)\,X(k) \tag{4.8}$$

Because interpolation is also a filtering operation, fast interpolation can also be performed in the frequency domain using the FFT. The next section describes the implemen-

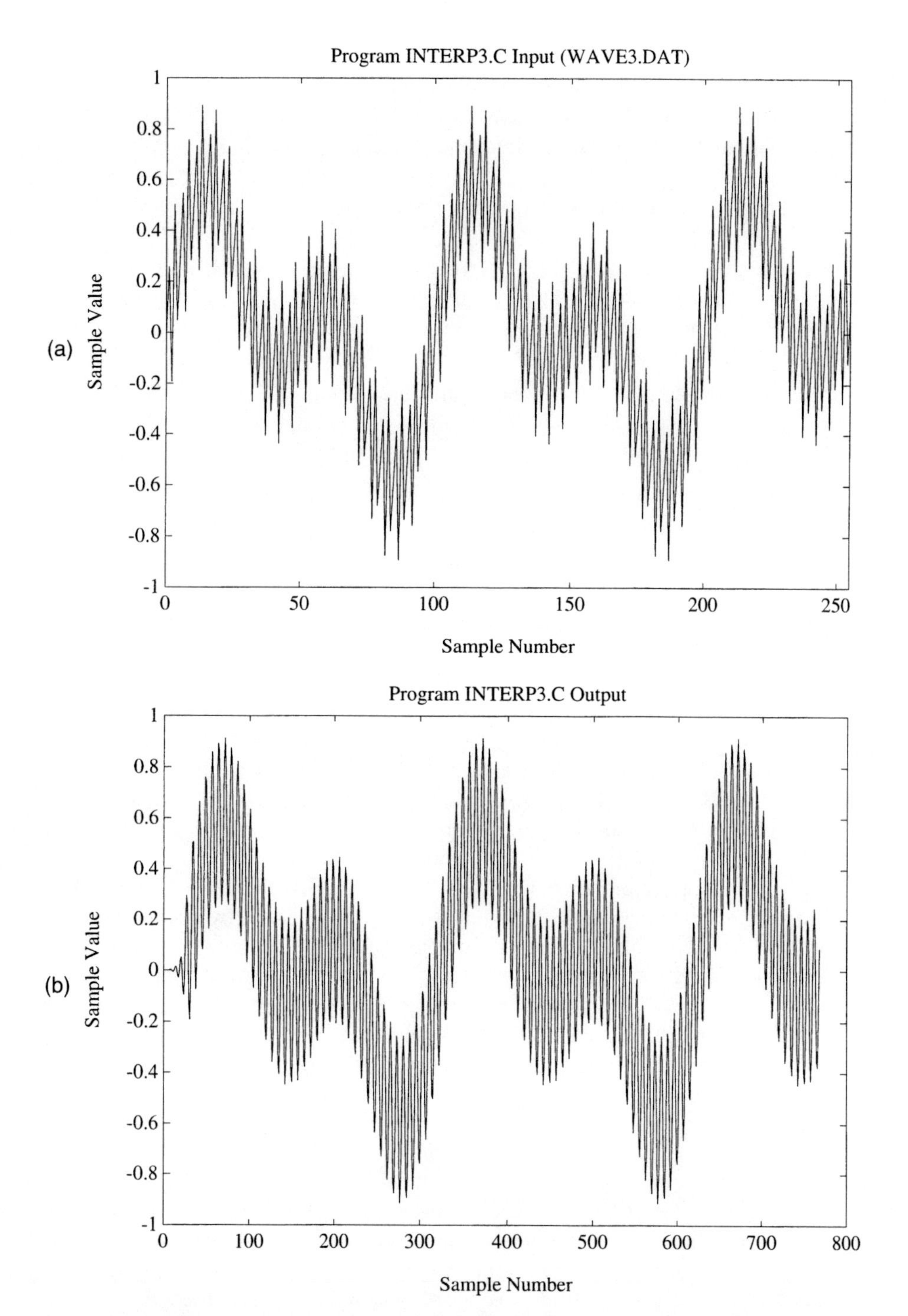

FIGURE 4.10 **(a)** Example of INTERP3 for 3:1 interpolation. Original WAVE3.DAT. **(b)** 3:1 interpolated WAVE3.DAT output.

tation of real-time filters using FFT fast convolution methods, and section 4.4.2 describes a real-time implementation of frequency domain interpolation.

4.4.1 Fast Convolution Using FFT Methods

Equation (4.8) indicates that if the frequency domain representations of $h(n)$ and $x(n)$ are known, then $Y(k)$ can be calculated by simple multiplication. The sequence $y(n)$ can then be obtained by inverse Fourier transform. This sequence of steps is detailed below:

(1) Create the array $H(k)$ from the impulse response $h(n)$ using the FFT.

(2) Create the array $X(k)$ from the sequence $x(n)$ using the FFT.

(3) Multiply H by X point by point thereby obtaining $Y(k)$.

(4) Apply the inverse FFT to $Y(k)$ in order to create $y(n)$.

There are several points to note about this procedure. First, very often the impulse response $h(n)$ of the filter does not change over many computations of the convolution equation. Therefore, the array $H(k)$ need only be computed once and can be used repeatedly, saving a large part of the computation burden of the algorithm.

Second, it must be noted that $h(n)$ and $x(n)$ may have different lengths. In this case, it is necessary to create two equal length sequences by adding zero-value samples at the end of the shorter of the two sequences. This is commonly called *zero filling* or *zero padding*. This is necessary because all FFT lengths in the procedure must be equal. Also, when using the radix 2 FFT all sequences to be processed must have a power of 2 length. This can require zero filling of both sequences to bring them up to the next higher value that is a power of 2.

Finally, in order to minimize circular convolution edge effects (the distortions that occur at computation points where each value of $h(n)$ does not have a matching value in $x(n)$ for multiplication), the length of $x(n)$ is often extended by the original length of $h(n)$ by adding zero values to the end of the sequence. The problem can be visualized by thinking of the convolution equation as a process of sliding a short sequence, $h(n)$, across a longer sequence, $x(n)$, and taking the sum of products at each translation point. As this translation reaches the end of the $x(n)$ sequence, there will be sums where not all $h(n)$ values match with a corresponding $x(n)$ for multiplication. At this point the output $y(n)$ is actually calculated using points from the beginning of $x(n)$, which may not be as useful as at the other central points in the convolution. This circular convolution effect cannot be avoided when using the FFT for fast convolution, but by zero filling the sequence its results are made predictable and repeatable.

The speed of the FFT makes convolution using the Fourier transform a practical technique. In fact, in many applications fast convolution using the FFT can be significantly faster than normal time domain convolution. As with other FFT applications, there is less advantage with shorter sequences and with very small lengths the overhead can create a penalty. The number of real multiply/accumulate operations required for fast convolution of an N length input sequence (where N is a large number, a power of 2 and real FFTs are used) with a fixed filter sequence is $2*N*[1 + 2*\log_2(N)]$. For example, when N is 1,024 and M is 100, fast convolution is as much as 2.15 times faster.

The program RFAST (see Listing 4.10) illustrates the use of the **fft** function for fast convolution (see Listing 4.11 for a C language implementation). Note that the inverse FFT is performed by swapping the real and imaginary parts of the input and output of the **fft** function. The overlap and save method is used to filter the continuous real-time input and generate a continuous output from the 1024 point FFT. The convolution problem is filtering with the 35-tap low pass FIR filter as was used in section 4.2.2. The filter is defined in the FILTER.H header file (variable **fir_lpf35**). The RFAST program can be used to generate results similar to the result shown in Figure 4.7(b).

(text continues on page 176)

```
#include <stdlib.h>
#include <stdio.h>
#include <string.h>
#include <math.h>
#include "rtdspc.h"
#include "filter.h"

/***********************************************************************

RFAST.C - Realtime fast convolution using the FFT

This program performs fast convolution using the FFT. It performs
the convolution required to implement a 35 point FIR filter
(stored in variable fir_lpf35) on an
arbitrary length realtime input. The filter is
a LPF with 40 dB out of band rejection. The 3 dB point is at a
relative frequency of approximately .25*fs.

***********************************************************************/

/* FFT length must be a power of 2 */
#define FFT_LENGTH 1024
#define M 10              /* must be log2(FFT_LENGTH) */
#define FILTER_LENGTH 35

void main()
{
 int           i, j;
 float         tempflt;
 COMPLEX       *samp, *filt;
 static float input_save[FILTER_LENGTH];

/* power of 2 length of FFT and complex allocation */
  samp = (COMPLEX *) calloc(FFT_LENGTH, sizeof(COMPLEX));
  if(!samp){
```

LISTING 4.10 Program RFAST to perform real-time fast convolution using the overlap and save method. (*Continued*)

```
    exit(1);
 }

/* Zero fill the filter to the sequence length */
 filt = (COMPLEX *) calloc(FFT_LENGTH, sizeof(COMPLEX));
 if(!filt){
    exit(1);
 }

/* copy the filter into complex array and scale by 1/N for inverse FFT */
 tempflt = 1.0/FFT_LENGTH;
 for(i = 0 ; i < FILTER_LENGTH ; i++)
     filt[i].real = tempflt*fir_lpf35[i];

/* FFT the zero filled filter impulse response */
 fft(filt,M);

/* read in one FFT worth of samples to start, imag already zero */
 for(i = 0 ; i < FFT_LENGTH-FILTER_LENGTH ; i++)
    samp[i].real = getinput();

/* save the last FILTER_LENGTH points for next time */
 for(j = 0 ; j < FILTER_LENGTH ; j++, i++)
    input_save[j] = samp[i].real = getinput();

 while(1) {

/* do FFT of samples */
    fft(samp,M);

/* Multiply the two transformed sequences */
/* swap the real and imag outputs to allow a forward FFT instead of
inverse FFT */
    for(i = 0 ; i < FFT_LENGTH ; i++) {
      tempflt = samp[i].real * filt[i].real
                          - samp[i].imag * filt[i].imag;
      samp[i].real = samp[i].real * filt[i].imag
                          + samp[i].imag * filt[i].real;
      samp[i].imag = tempflt;
    }

/* Inverse fft the multiplied sequences */
    fft(samp,M);

/* Write the result out to a dsp data file */
/* because a forward FFT was used for the inverse FFT,
```

LISTING 4.10 (*Continued*)

```
 the output is in the imag part */
     for(i = FILTER_LENGTH ; i < FFT_LENGTH ; i++) sendout(samp[i].imag);

 /* overlap the last FILTER_LENGTH-1 input data points in the next FFT */
     for(i = 0; i < FILTER_LENGTH ; i++) {
       samp[i].real = input_save[i];
       samp[i].imag = 0.0;
     }

     for( ; i < FFT_LENGTH-FILTER_LENGTH ; i++) {
       samp[i].real = getinput();
       samp[i].imag = 0.0;
     }

 /* save the last FILTER_LENGTH points for next time */
     for(j = 0 ; j < FILTER_LENGTH ; j++, i++) {
       input_save[j] = samp[i].real = getinput();
       samp[i].imag = 0.0;
     }
  }
 }
```

LISTING 4.10 (*Continued*)

```
/**************************************************************************

fft - In-place radix 2 decimation in time FFT

Requires pointer to complex array, x and power of 2 size of FFT, m
(size of FFT = 2**m). Places FFT output on top of input COMPLEX array.

void fft(COMPLEX *x, int m)

*************************************************************************/

void fft(COMPLEX *x,int m)
{
    static COMPLEX *w;           /* used to store the w complex array */
    static int mstore = 0;       /* stores m for future reference */
    static int n = 1;            /* length of fft stored for future */

    COMPLEX u,temp,tm;
    COMPLEX *xi,*xip,*xj,*wptr;

    int i,j,k,l,le,windex;

    double arg,w_real,w_imag,wrecur_real,wrecur_imag,wtemp_real;
```

LISTING 4.11 Radix 2 FFT function **fft(x,m)**. (*Continued*)

```
    if(m != mstore) {

/* free previously allocated storage and set new m */

        if(mstore != 0) free(w);
        mstore = m;
        if(m == 0) return;          /* if m=0 then done */

/* n = 2**m = fft length */

        n = 1 << m;
        le = n/2;

/* allocate the storage for w */

        w = (COMPLEX *) calloc(le-1,sizeof(COMPLEX));
        if(!w) {
            exit(1);
        }

/* calculate the w values recursively */

        arg = 4.0*atan(1.0)/le;         /* PI/le calculation */
        wrecur_real = w_real = cos(arg);
        wrecur_imag = w_imag = -sin(arg);
        xj = w;
        for (j = 1 ; j < le ; j++) {
            xj->real = (float)wrecur_real;
            xj->imag = (float)wrecur_imag;
            xj++;
            wtemp_real = wrecur_real*w_real - wrecur_imag*w_imag;
            wrecur_imag = wrecur_real*w_imag + wrecur_imag*w_real;
            wrecur_real = wtemp_real;
        }
    }

/* start fft */

    le = n;
    windex = 1;
    for (l = 0 ; l < m ; l++) {
        le = le/2;

/* first iteration with no multiplies */

        for(i = 0 ; i < n ; i = i + 2*le) {
            xi = x + i;
            xip = xi + le;
```

LISTING 4.11 *(Continued)*

```
            temp.real = xi->real + xip->real;
            temp.imag = xi->imag + xip->imag;
            xip->real = xi->real - xip->real;
            xip->imag = xi->imag - xip->imag;
            *xi = temp;
        }

/* remaining iterations use stored w */

        wptr = w + windex - 1;
        for (j = 1 ; j < le ; j++) {
            u = *wptr;
            for (i = j ; i < n ; i = i + 2*le) {
                xi = x + i;
                xip = xi + le;
                temp.real = xi->real + xip->real;
                temp.imag = xi->imag + xip->imag;
                tm.real = xi->real - xip->real;
                tm.imag = xi->imag - xip->imag;
                xip->real = tm.real*u.real - tm.imag*u.imag;
                xip->imag = tm.real*u.imag + tm.imag*u.real;
                *xi = temp;
            }
            wptr = wptr + windex;
        }
        windex = 2*windex;
    }

/* rearrange data by bit reversing */

    j = 0;
    for (i = 1 ; i < (n-1) ; i++) {
        k = n/2;
        while(k <= j) {
            j = j - k;
            k = k/2;
        }
        j = j + k;
        if (i < j) {
            xi = x + i;
            xj = x + j;
            temp = *xj;
            *xj = *xi;
            *xi = temp;
        }
    }
}
```

LISTING 4.11 *(Continued)*

4.4.2 Interpolation Using the FFT

In section 4.3.2 time domain interpolation was discussed and demonstrated using several short FIR filters. In this section, the same process is demonstrated using FFT techniques. The steps involved in 2:1 interpolation using the FFT are as follows:

(1) Perform an FFT with a power of 2 length (N) which is greater than or equal to the length of the input sequence.

(2) Zero pad the frequency domain representation of the signal (a complex array) by inserting $N - 1$ zeros between the positive and negative half of the spectrum. The Nyquist frequency sample output of the FFT (at the index $N/2$) is divided by 2 and placed with the positive and negative parts of the spectrum, this results in a symmetrical spectrum for a real input signal.

(3) Perform an inverse FFT with a length of $2N$.

(4) Multiply the interpolated result by a factor of 2 and copy the desired portion of the result that represents the interpolated input, this is all the inverse FFT samples if the input length was a power of 2.

Listing 4.12 shows the program INTFFT2.C that performs 2:1 interpolation using the above procedure and the **fft** function (shown in Listing 4.11). Note that the inverse FFT is performed by swapping the real and imaginary parts of the input and output of the **fft** function. Figure 4.11 shows the result of using the INTFFT2 program on the 128 samples of the WAVE3.DAT input file used in the previous examples in this chapter (these 256 samples are shown in detail in Figure 4.10(a)). Note that the output length is twice as large (512) and more of the sine wave nature of the waveform can be seen in the interpolated result. The INTFFT2 program can be modified to interpolate by a larger power of 2 by increasing the number of zeros added in step (2) listed above. Also, because the FFT is employed, frequencies as high as the Nyquist rate can be accurately interpolated. FIR filter interpolation has a upper frequency limit because of the frequency response of the filter (see section 4.3.1).

```
#include <stdlib.h>
#include <stdio.h>
#include <string.h>
#include <math.h>
#include "rtdspc.h"

/**************************************************************************

INTFFT2.C - Interpolate 2:1 using FFT

Generates 2:1 interpolated time domain data.

*************************************************************************/
```

LISTING 4.12 Program INTFFT2.C used to perform 2:1 interpolation using the FFT. (*Continued*)

```
#define LENGTH 256
#define M 8                /* must be log2(FFT_LENGTH) */

main()
{
    int         i;
    float       temp;
    COMPLEX     *samp;

/* allocate the complex array (twice as long) */
    samp = (COMPLEX *) calloc(2*LENGTH, sizeof(COMPLEX));
    if(!samp) {
        printf("\nError allocating fft memory\n");
        exit(1);
    }

/* copy input signal to complex array and do the fft */
    for (i = 0; i < LENGTH; i++) samp[i].real = getinput();

    fft(samp,M);

/* swap the real and imag to do the inverse fft */
    for (i = 0; i < LENGTH; i++) {
        temp = samp[i].real;
        samp[i].real = samp[i].imag;
        samp[i].imag = temp;
    }

/* divide the middle frequency component by 2 */
    samp[LENGTH/2].real = 0.5*samp[LENGTH/2].real;
    samp[LENGTH/2].imag = 0.5*samp[LENGTH/2].imag;

/* zero pad and move the negative frequencies */
    samp[3*LENGTH/2] = samp[LENGTH/2];
    for (i = LENGTH/2 + 1; i < LENGTH ; i++) {
        samp[i+LENGTH] = samp[i];
        samp[i].real = 0.0;
        samp[i].imag = 0.0;
    }

/* do inverse fft by swapping input and output real & imag */
    fft(samp,M+1);

/* copy to output and multiply by 2/(2*LENGTH) */
    temp = 1.0/LENGTH;
    for (i=0; i < 2*LENGTH; i++) sendout(temp*samp[i].imag);
}
```

LISTING 4.12 *(Continued)*

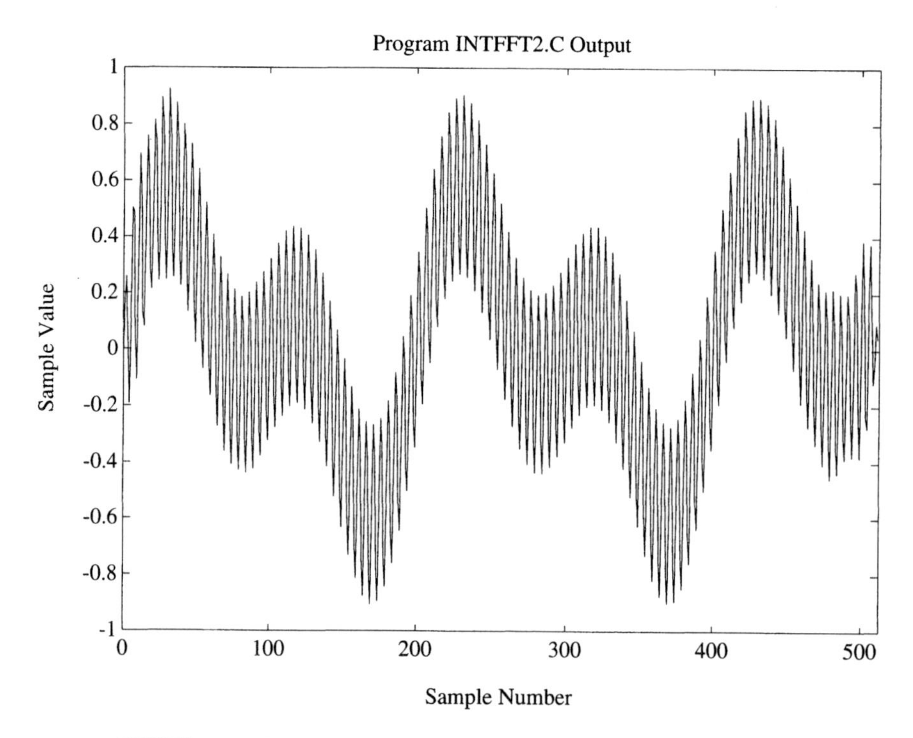

FIGURE 4.11 Example use of the INTFFT2 program used to interpolate WAVE3 signal by a 2:1 ratio.

4.5 OSCILLATORS AND WAVEFORM SYNTHESIS

The generation of pure tones is often used to synthesize new sounds in music or for testing DSP systems. The basic oscillator is a special case of an IIR filter where the poles are on the unit circle and the initial conditions are such that the input is an impulse. If the poles are moved outside the unit circle, the oscillator output will grow at an exponential rate. If the poles are placed inside the unit the circle, the output will decay toward zero. The state (or history) of the second order section determines the amplitude and phase of the future output. The next section describes the details of this type of oscillator. Section 4.5.2 considers another method to generate periodic waveforms of different frequencies —the wave table method. In this case any period waveform can be used to generate a fundamental frequency with many associated harmonics.

4.5.1 IIR Filters as Oscillators

The impulse response of a continuous time second order oscillator is given by

$$y(t) = e^{-dt} \frac{\sin(\omega t)}{\omega}. \tag{4.9}$$

If $d > 0$ then the output will decay toward zero and the peak will occur at

$$t_{peak} = \frac{\tan^{-1}(\omega/d)}{\omega}. \tag{4.10}$$

The peak value will be

$$y(t_{peak}) = \frac{e^{-dt_{peak}}}{\sqrt{d^2 + \omega^2}}. \tag{4.11}$$

A second-order difference can be used to generate a response that is an approximation of this continuous time output. The equation for a second-order discrete time oscillator is based on an IIR filter and is as follows:

$$y_{n+1} = c_1 y_n - c_2 y_{n-1} + b_1 x_n, \tag{4.12}$$

where the x input is only present for $t = 0$ as an initial condition to start the oscillator and

$$c_1 = 2e^{-d\tau}\cos(\omega\tau)$$
$$c_2 = e^{-2d\tau}$$

where τ is the sampling period ($1/f_s$) and ω is 2π times the oscillator frequency.

The frequency and rate of change of the envelope of the oscillator output can be changed by modifying the values of d and ω on a sample by sample basis. This is illustrated in the OSC program shown in Listing 4.13. The output waveform grows from a peak value of 1.0 to a peak value of 16000 at sample number 5000. After sample 5000 the envelope of the output decays toward zero and the frequency is reduced in steps every 1000 samples. A short example output waveform is shown in Figure 4.12.

4.5.2 Table-Generated Waveforms

Listing 4.14 shows the program WAVETAB.C, which generates a fundamental frequency at a particular musical note given by the variable **key**. The frequency in Hertz is related to the integer **key** as follows:

$$f = 440 \bullet 2^{key/12} \tag{4.13}$$

Thus, a **key** value of zero will give 440 Hz, which is the musical note A above middle C. The WAVETAB.C program starts at a key value of –24 (two octaves below A) and steps through a chromatic scale to key value 48 (4 octaves above A). Each sample output value is calculated using a linear interpolation of the 300 values in the table **gwave**. The 300 sample values are shown in Figure 4.13 as an example waveform. The **gwave** array is 301 elements to make the interpolation more efficient. The first element (0) and the last element (300) are the same, creating a circular interpolated waveform. Any waveform can be substituted to create different sounds. The amplitude of the output is controlled by the **env** variable, and grows and decays at a rate determined by **trel** and **amp** arrays.

```
#include <stdlib.h>
#include <stdio.h>
#include <math.h>
#include "rtdspc.h"

    float osc(float,float,int);
    float rate,freq;
    float amp = 16000;

void main()
{
    long int i,length = 100000;

/* calculate the rate required to get to desired amp in 5000 samples */
    rate = (float)exp(log(amp)/(5000.0));

/* start at 4000 Hz */
    freq = 4000.0;

/* first call to start up oscillator */
    sendout(osc(freq,rate,-1));
/* special case for first 5000 samples to increase amplitude */
    for(i = 0 ; i < 5000 ; i++)
        sendout(osc(freq,rate,0));

/* decay the osc 10% every 5000 samples */
    rate = (float)exp(log(0.9)/(5000.0));

    for( ; i < length ; i++) {
        if((i%1000) == 0) {          /* change freq every 1000 samples */
            freq = 0.98*freq;
            sendout(osc(freq,rate,1));
        }
        else {                        /* normal case */
            sendout(osc(freq,rate,0));
        }
    }
    flush();
}

/* Function to generate samples from a second order oscillator
        rate = envelope rate of change parameter (close to 1).
    change_flag = indicates that frequency and/or rate have changed.
*/

float osc(float freq,float rate,int change_flag)
{
/* calculate this as a static so it never happens again */
    static float two_pi_div_sample_rate = (float)(2.0 * PI / SAMPLE_RATE);
    static float y1,y0,a,b,arg;
    float out,wosc;
```

LISTING 4.13 Program OSC to generate a sine wave signal with a variable frequency and envelope using a second-order IIR section. (*Continued*)

```
/* change_flag:
    -1 = start new sequence from t=0
     0 = no change, generate next sample in sequence
     1 = change rate or frequency after start
*/
    if(change_flag != 0) {
/* assume rate and freq change every time */
        wosc = freq * two_pi_div_sample_rate;
        arg = 2.0 * cos(wosc);
        a = arg * rate;
        b = -rate * rate;

        if(change_flag < 0) {     /* re-start case, set state variables */
            y0 = 0.0f;
            return(y1 = rate*sin(wosc));
        }
    }
/* make new sample */
    out = a*y1 + b*y0;
    y0 = y1;
    y1 = out;
    return(out);
}
```

LISTING 4.13 (*Continued*)

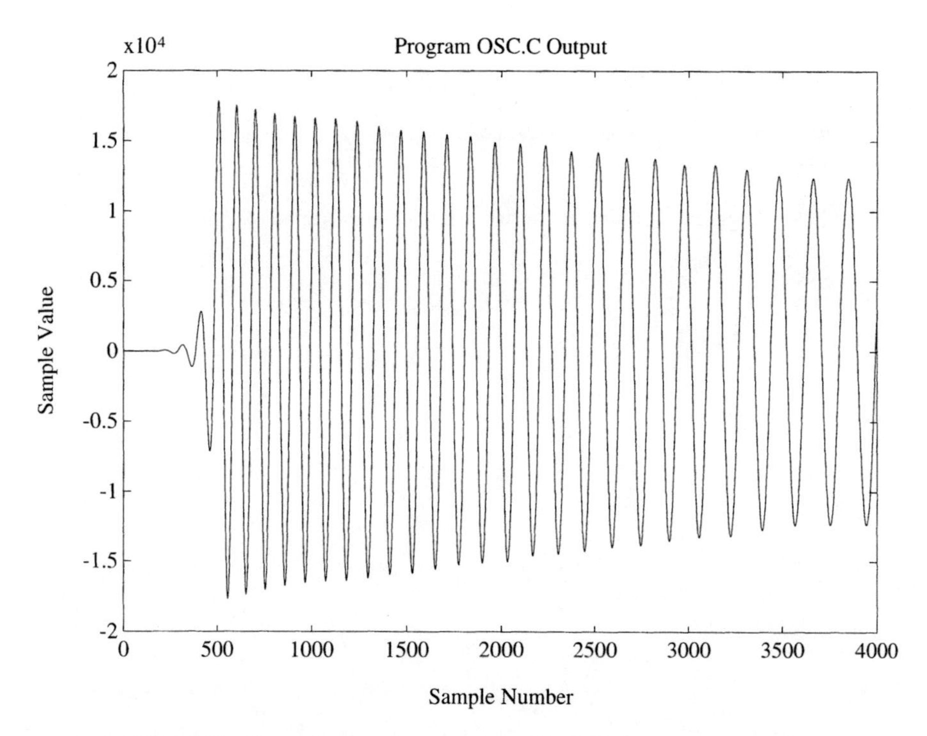

FIGURE 4.12 Example signal output from the OSC.C program (modified to reach peak amplitude in 500 samples and change frequency every 500 sample for display purposes).

```
#include <stdlib.h>
#include <math.h>
#include "rtdspc.h"
#include "gwave.h"         /* gwave[301] array */

/* Wavetable Music Generator 4-20-94 PME */

    int key;

void main()
{
    int t,told,ci,k;
    float ampold,rate,env,wave_size,dec,phase,frac,delta,sample;
    register long int i,endi;
    register float sig_out;

    static float trel[5] = {     0.02 ,     0.14,      0.6, 1.0, 0.0 };
    static float amps[5] = { 15000.0 , 10000.0, 4000.0, 10.0, 0.0 };
    static float rates[10];
    static int tbreaks[10];

    wave_size = 300.0;      /* dimension of original wave */
    endi = 96000;             /* 2 second notes */

    for(key = -24 ; key < 48 ; key++) {

/* decimation ratio for key semitones down */
        dec = powf(2.0,0.0833333333*(float)key);

/* calculate the rates required to get the desired amps */
        i = 0;
        told = 0;
        ampold = 1.0;          /* always starts at unity */
        while(amps[i] > 1.0) {
            t = trel[i]*endi;
            rates[i] = expf(logf(amps[i]/ampold)/(t-told));
            ampold = amps[i];
            tbreaks[i] = told = t;
            i++;
        }

        phase = 0.0;
        rate = rates[0];
        env = 1.0;
        ci = 0;
        for(i = 0 ; i < endi ; i++) {
/* calculate envelope amplitude */
```

LISTING 4.14 Program WAVETAB to generate periodic waveform at any frequency. (*Continued*)

```
            if(i == tbreaks[ci]) rate = rates[++ci];
            env = rate*env;
/* determine interpolated sample value from table */
            k = (int)phase;
            frac = phase - (float)k;
            sample = gwave[k];
            delta = gwave[k+1] - sample; /* possible wave_size+1 access */
            sample += frac*delta;
/* calculate output and send to DAC */
            sig_out = env*sample;
            sendout(sig_out);
/* calculate next phase value */
            phase += dec;
            if(phase >= wave_size) phase -= wave_size;
        }
    }
    flush();
}
```

LISTING 4.14 (*Continued*)

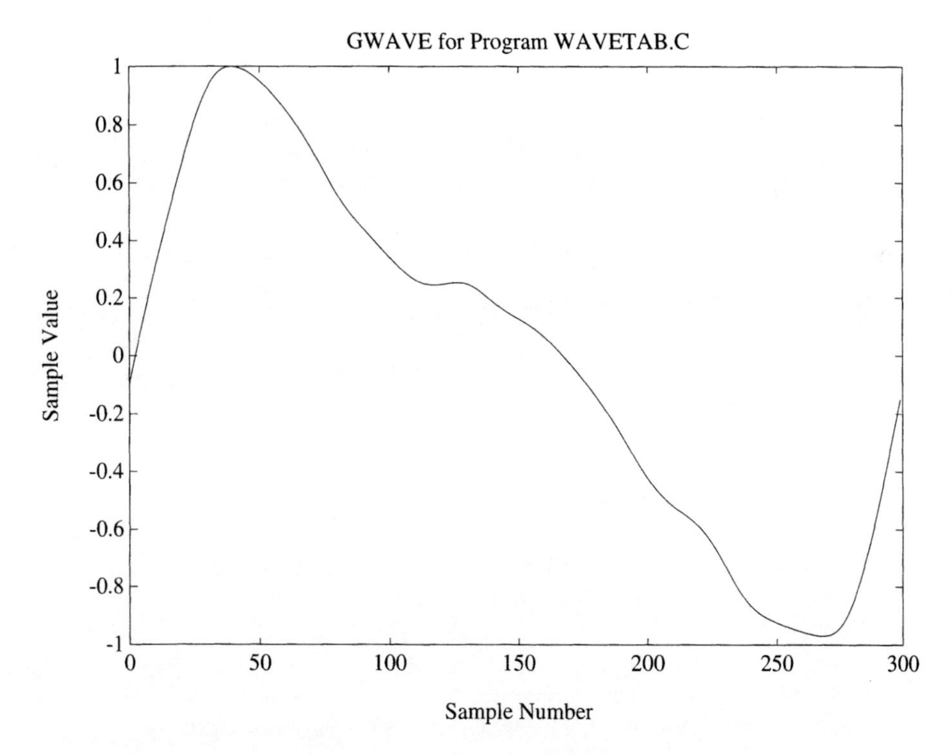

FIGURE 4.13 Example waveform (`gwave[301]` array) used by program WAVETAB.

4.6 REFERENCES

ANTONIOU, A. (1979). *Digital Fitters: Analysis and Design.* New York: McGraw-Hill.

BRIGHAM, E.O. (1988). *The Far Fourier Transform and Its Applications.* Englewood Cliffs, NJ: Prentice Hall.

CROCHIERE, R.E. and RABINER, L.R. (1983). *Multirate Digital Signal Processing.* Englewood Cliffs, NJ: Prentice Hall.

ELIOTT, D.F. (Ed.). (1987). *Handbook of Digital Signal Processing.* San Diego, CA: Academic Press.

EMBREE, P. and KIMBLE B. (1991). *C Language Algorithms for Digital Signal Processing.* Englewood Cliffs, NJ: Prentice Hall.

GHAUSI, M.S. and LAKER, K.R. (1981). *Modern Filter Design: Active RC and Switched Capacitor.* Englewood Cliffs, NJ: Prentice Hall.

IEEE DIGITAL SIGNAL PROCESSING COMMITTEE (Ed.). (1979). *Programs for Digital Signal Processing.* New York: IEEE Press.

JOHNSON, D.E., JOHNSON, J.R. and MOORE, H.P. (1980). *A Handbook of Active Filters.* Englewood Cliffs, NJ: Prentice Hall.

JONG, M.T. (1992). *Methods of Discrete Signal and System Analysis.* New York: McGraw-Hill.

KAISER, J. F. and SCHAFER, R. W. (Feb. 1980). On the Use of the /0-Sinh Window for Spectrum Analysis. *IEEE Transactions on Acoustics, Speech, and Signal Processing, (ASSP-28)* (1), 105–107.

KNUTH, D.E. (1981). *Seminumerical Algorithms, The Art of Computer Programming, Vol. 2.* (2nd ed.). Reading, MA: Addison-Wesley.

McCLELLAN, J., PARKS, T. and RABINER, L.R. (1973). A Computer Program for Designing Optimum FIR Linear Phase Digital Filters. *IEEE Transactions on Audio and Electro-acoustics, AU-21.* (6), 506–526.

MOLER, C., LITTLE, J. and BANGERT, S. (1987). *PC-MATLAB User's Guide.* Natick, MA: The Math Works.

MOSCHUYTZ, G.S. and HORN, P. (1981). *Active Filter Design Handbook.* New York: John Wiley & Sons.

OPPENHEIM, A. and SCHAFER, R. (1975). *Digital Signal Processing,* Englewood Cliffs, NJ: Prentice Hall.

OPPENHEIM, A. and SCHAFER, R. (1989). *Discrete-time Signal Processing.* Englewood Cliffs, NJ: Prentice Hall.

PAPOULIS, A. (1984). *Probability, Random Variables and Stochastic Processes,* (2nd ed.). New York: McGraw-Hill.

PARK, S.K. and MILLER, K.W. (Oct. 1988). Random Number Generators: Good Ones Are Hard to Find. *Communications of the ACM, (31)* (10).

PARKS, T.W. and BURRUS, C.S. (1987). *Digital Filter Design.* New York: John Wiley & Sons.

PRESS W.H., FLANNERY, B.P., TEUKOLSKY, S.A. and VETTERLING, W.T. (1987). *Numerical Recipes.* New York: Cambridge Press.

RABINER, L. and GOLD, B. (1975). *Theory and Application of Digital Signal Processing.* Englewood Cliffs, NJ: Prentice Hall.

STEARNS, S. and DAVID, R. (1988). *Signal Processing Algorithms.* Englewood Cliffs, NJ: Prentice Hall.

VAN VALKENBURG, M.E. (1982). *Analog Filter Design.* New York: Holt, Rinehart and Winston.

ZVEREV, A. I. (1967). *Handbook of Filter Synthesis.* New York: John Wiley & Sons.

CHAPTER 5

REAL-TIME DSP APPLICATIONS

This chapter combines the DSP principles described in the previous chapters with the specifications of real-time systems designed to solve real-world problems and provide complete software solutions for several DSP applications. Applications of FFT spectrum analysis are described in section 5.1. Speech and music processing are considered in sections 5.3 and 5.4. Adaptive signal processing methods are illustrated in section 5.2 (parametric signal modeling) and section 5.5 (adaptive frequency tracking).

5.1 FFT POWER SPECTRUM ESTIMATION

Signals found in most practical DSP systems do not have a constant power spectrum. The spectrum of radar signals, communication signals, and voice waveforms change continually with time. This means that the FFT of a single set of samples is of very limited use. More often a series of spectra are required at time intervals determined by the type of signal and information to be extracted.

Power spectral estimation using FFTs provides these power spectrum snapshots (called *periodograms*). The average of a series of periodograms of the signal is used as the estimate of the spectrum of the signal at a particular time. The parameters of the *average periodogram spectral estimate* are:

(1) Sample rate: Determines maximum frequency to be estimated

(2) Length of FFT: Determines the resolution (smallest frequency difference detectable)

(3) Window: Determines the amount of spectral leakage and affects resolution and noise floor

(4) Amount of overlap between successive spectra: Determines accuracy of the estimate, directly affects computation time

(5) Number of spectra averaged: Determines maximum rate of change of the detectable spectra and directly affects the noise floor of the estimate

5.1.1 Speech Spectrum Analysis

One of the common application areas for power spectral estimation is speech processing. The power spectra of a voice signal give essential clues to the sound being made by the speaker. Almost all the information in voice signals is contained in frequencies below 3,500 Hz. A common voice sampling frequency that gives some margin above the Nyquist rate is 8,000 Hz. The spectrum of a typical voice signal changes significantly every 10 msec or 80 samples at 8,000 Hz. As a result, popular FFT sizes for speech processing are 64 and 128 points.

Included on the MS-DOS disk with this book is a file called CHKL.TXT. This is the recorded voice of the author saying the words "chicken little." These sounds were chosen because of the range of interesting spectra that they produced. By looking at a plot of the CHKL.TXT samples (see Figure 5.1) the break between words can be seen and the

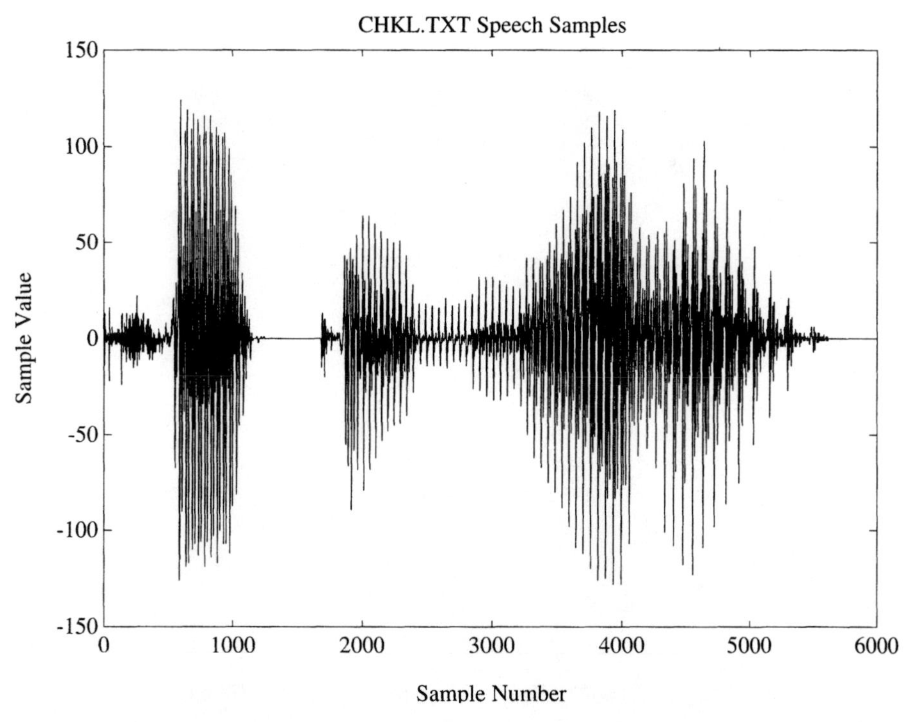

FIGURE 5.1 Original CHKL.TXT data file consisting of the author's words "chicken little" sampled at 8 kHz (6000 samples are shown).

relative volume can be inferred from the envelope of the waveform. The frequency content is more difficult to determine from this plot.

The program RTPSE (see Listing 5.1) accepts continuous input samples (using **getinput()**) and generates a continuous set of spectral estimates. The power spectral estimation parameters, such as FFT length, overlap, and number of spectra averaged, are set by the program to default values. The amount of overlap and averaging can be changed in real-time. RTPSE produces an output consisting of a spectral estimate every 4 input samples. Each power spectral estimate is the average spectrum of the input file over the past 128 samples (16 FFT outputs are averaged together).

Figure 5.2 shows a contour plot of the resulting spectra plotted as a frequency versus time plot with the amplitude of the spectrum indicated by the contours. The high fre-

```
#include <stdlib.h>
#include <stdio.h>
#include <math.h>
#include "rtdspc.h"

/*********************************************************************

RTPSE.C - Real-Time Power spectral estimation using the FFT

This program does power spectral estimation on input samples.
The average power spectrum in each block is determined
and used to generate a series of outputs.

Length of each FFT snapshot: 64 points
Number of FFTs to average: 16 FFTs
Amount of overlap between each FFT: 60 points

*********************************************************************/

/* FFT length must be a power of 2 */
#define FFT_LENGTH 64
#define M 6              /* must be log2(FFT_LENGTH) */

/* these variables global so they can be changed in real-time */
  int numav = 16;
  int ovlap = 60;

main()
{
  int         i,j,k;
  float       scale,tempflt;
```

LISTING 5.1 Program RTPSE to perform real-time power spectral estimation using the FFT. (*Continued*)

```
  static float   mag[FFT_LENGTH], sig[FFT_LENGTH], hamw[FFT_LENGTH];
  static COMPLEX samp[FFT_LENGTH];

/* overall scale factor */
  scale = 1.0f/(float)FFT_LENGTH;
  scale *= scale/(float)numav;

/* calculate hamming window */
  tempflt = 8.0*atan(1.0)/(FFT_LENGTH-1);
  for(i = 0 ; i < FFT_LENGTH ; i++)
    hamw[i] = 0.54 - 0.46*cos(tempflt*i);

/* read in the first FFT_LENGTH samples, overlapped samples read in loop */
  for(i = 0 ; i < FFT_LENGTH ; i++) sig[i] = getinput();

  for(;;) {

    for (k=0; k<FFT_LENGTH; k++) mag[k] = 0;

    for (j=0; j<numav; j++){

      for (k=0; k<FFT_LENGTH; k++){
        samp[k].real = hamw[k]*sig[k];
        samp[k].imag = 0;
      }

      fft(samp,M);

      for (k=0; k<FFT_LENGTH; k++){
        tempflt  = samp[k].real * samp[k].real;
        tempflt += samp[k].imag * samp[k].imag;
        tempflt = scale*tempflt;
        mag[k] += tempflt;
      }

/* overlap the new samples with the old */
      for(k = 0 ; k < ovlap ; k++) sig[k] = sig[k+FFT_LENGTH-ovlap];
      for( ; k < FFT_LENGTH ; k++) sig[k] = getinput();
    }

/*  Take log after averaging the magnitudes.  */
    for (k=0; k<FFT_LENGTH/2; k++){
      tempflt = mag[k];
      if(tempflt < 1.e-10f) tempflt = 1.e-10f;
      sendout(10.0f*log10(tempflt));
    }
  }
}
```

LISTING 5.1 *(Continued)*

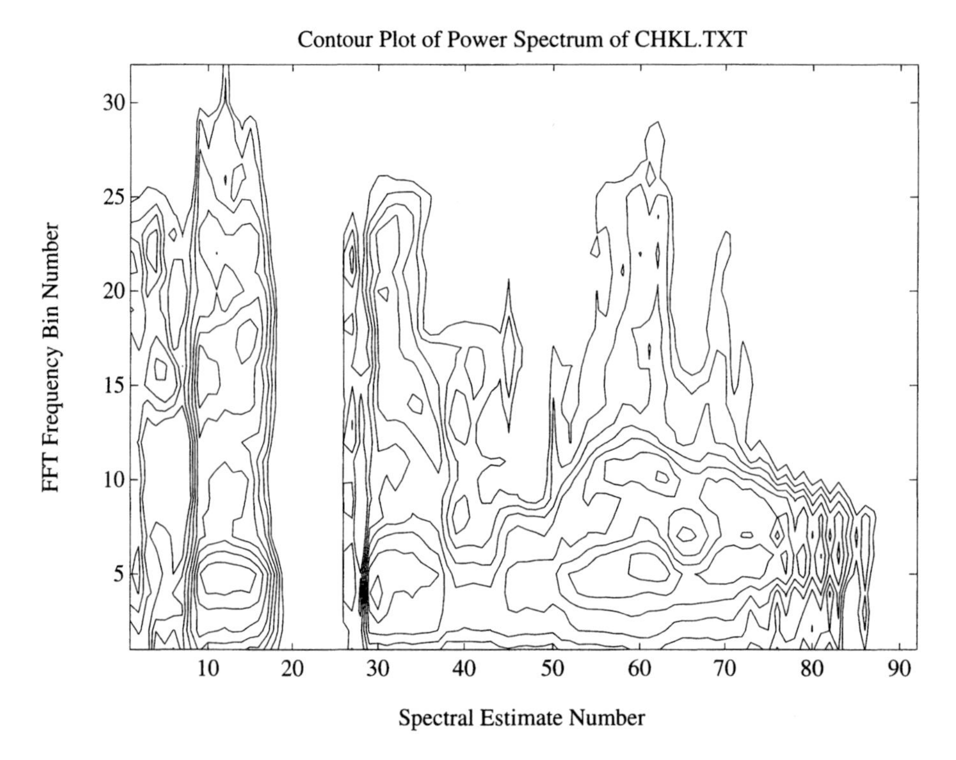

FIGURE 5.2 Contour plot of the power spectrum versus frequency and time obtained using the RTPSE program with the input file CHKL.TXT. Contours are at 5 dB intervals and the entire 2D power spectrum is normalized to 0 dB.

quency content of the "chi'' part of "chicken" and the lower frequency content of "little'' are clearly indicated.

5.1.2 Doppler Radar Processing

Radar signals are normally transmitted at a very high frequency (usually greater than 100 MHz), but with a relatively narrow bandwidth (several MHz). For this reason most radar signals are processed after mixing them down to baseband using a quadrature demodulator (see Skolnik, 1980). This gives a complex signal that can be processed digitally in real-time to produce a display for the radar operator. One type of display is a moving target indicator (MTI) display where moving targets are separated from stationary targets by signal processing. It is often important to know the speed and direction of the moving targets. Stationary targets are frequently encountered in radar because of fixed obstacles (antenna towers, buildings, trees) in the radar beam's path. The beam is not totally blocked by these targets, but the targets do return a large echo back to the receiver. These targets can be removed by determining their average amplitude from a series of echoes and subtracting them from each received echo. Any moving target will not be subtracted and can be further processed. A simple method to remove stationary echoes is to simply subtract successive echoes from each other (this is a simple highpass filter of the Doppler signal). The mean frequency of the remaining Doppler signal of the moving targets can then be determined using the complex FFT.

Listing 5.2 shows the program RADPROC.C, which performs the DSP required to remove stationary targets and then estimate the frequency of the remaining Doppler signal. In order to illustrate the operation of this program, the test data file RADAR.DAT was generated. These data represent the simulated received signal from a stationary target

```
#include <stdlib.h>
#include <math.h>
#include "rtdspc.h"

/***********************************************************************

RADPROC.C - Real-Time Radar processing

This program subtracts successive complex echo signals to
remove stationary targets from a radar signal and then
does power spectral estimation on the resulting samples.
The mean frequency is then estimated by finding the peak of the
FFT spectrum.

Requires complex input (stored real,imag) with 12
consecutive samples representing 12 range locations from
each echo.

***********************************************************************/

/* FFT length must be a power of 2 */
#define FFT_LENGTH 16
#define M 4              /* must be log2(FFT_LENGTH) */
#define ECHO_SIZE 12

void main()
{
  int            i,j,k;
  float          tempflt,rin,iin,p1,p2;
  static float   mag[FFT_LENGTH];
  static COMPLEX echos[ECHO_SIZE][FFT_LENGTH];
  static COMPLEX last_echo[ECHO_SIZE];
```

LISTING 5.2 Program RADPROC to perform real-time radar signal processing using the FFT. (*Continued*)

```
/* read in the first echo */
  for(i = 0 ; i < ECHO_SIZE ; i++) {
     last_echo[i].real = getinput();
     last_echo[i].imag = getinput();
  }

  for(;;) {
    for (j=0; j< FFT_LENGTH; j++){

/* remove stationary targets by subtracting pairs (highpass filter) */
      for (k=0; k< ECHO_SIZE; k++){
        rin = getinput();
        iin = getinput();
        echos[k][j].real = rin - last_echo[k].real;
        echos[k][j].imag = iin - last_echo[k].imag;
        last_echo[k].real = rin;
        last_echo[k].imag = iin;
      }
    }
/* do FFTs on each range sample */
    for (k=0; k< ECHO_SIZE; k++) {

      fft(echos[k],M);

      for(j = 0 ; j < FFT_LENGTH ; j++) {
        tempflt  = echos[k][j].real * echos[k][j].real;
        tempflt += echos[k][j].imag * echos[k][j].imag;
        mag[j] = tempflt;
      }
/* find the biggest magnitude spectral bin and output */
      tempflt = mag[0];
      i=0;
      for(j = 1 ; j < FFT_LENGTH ; j++) {
        if(mag[j] > tempflt) {
          tempflt = mag[j];
          i=j;
        }
      }
/* interpolate the peak loacation */
      p1 = mag[i] - mag[i-1];
      p2 = mag[i] - mag[i+1];
      sendout((float)i + (p1-p2)/(2*(p1+p2+1e-30)));
    }
  }
}
```

LISTING 5.2 *(Continued)*

added to a moving target signal with Gaussian noise. The data is actually a 2D matrix representing 12 consecutive complex samples (real,imag) along the echo in time (representing 12 consecutive range locations) with each of 33 echoes following one after another. The sampling rates and target speeds are not important to the illustration of the program. The output of the program is the peak frequency location from the 16-point FFT in bins (0 to 8 are positive frequencies and 9 to 15 are –7 to –1 negative frequency bins). A simple (and efficient) parabolic interpolation is used to give a fractional output in the results. The output from the RADPROC program using the RADAR.DAT as input is 24 consecutive numbers with a mean value of 11 and a small standard deviation due to the added noise. The first 12 numbers are from the first set of 16 echoes and the last 12 numbers are from the remaining echoes.

5.2 PARAMETRIC SPECTRAL ESTIMATION

The parametric approach to spectral estimation attempts to describe a signal as a result from a simple system model with a random process as input. The result of the estimator is a small number of parameters that completely characterize the system model. If the model is a good choice, then the spectrum of the signal model and the spectrum from other spectral estimators should be similar. The most common parametric spectral estimation models are based on AR, MA, or ARMA random process models as discussed in section 1.6.6 of chapter 1. Two simple applications of these models are presented in the next two sections.

5.2.1 ARMA Modeling of Signals

Figure 5.3 shows the block diagram of a system modeling problem that will be used to illustrate the adaptive IIR LMS algorithm discussed in detail in section 1.7.2 of chapter 1. Listing 5.3 shows the main program ARMA.C, which first filters white noise (generated using the Gaussian noise generator described in section 4.2.1 of chapter 4) using a second-order IIR filter, and then uses the LMS algorithm to adaptively determine the filter function.

Listing 5.4 shows the function **iir_biquad**, which is used to filter the white noise, and Listing 5.5 shows the adaptive filter function, which implements the LMS algorithm in a way compatible with real-time input. Although this is a simple ideal example where exact convergence can be obtained, this type of adaptive system can also be used to model more complicated systems, such as communication channels or control systems. The white noise generator can be considered a training sequence which is known to the algorithm; the algorithm must determine the transfer function of the system. Figure 5.4 shows the error function for the first 7000 samples of the adaptive process. The error reduces relatively slowly due to the poles and zeros that must be determined. FIR LMS algorithms generally converge much faster when the system can be modeled as a MA system (see section 5.5.2 for an FIR LMS example). Figure 5.5 shows the path of the pole coefficients (b0,b1) as they adapt to the final result where b0 = 0.748 and b1 = –0.272.

(text continues on page 198)

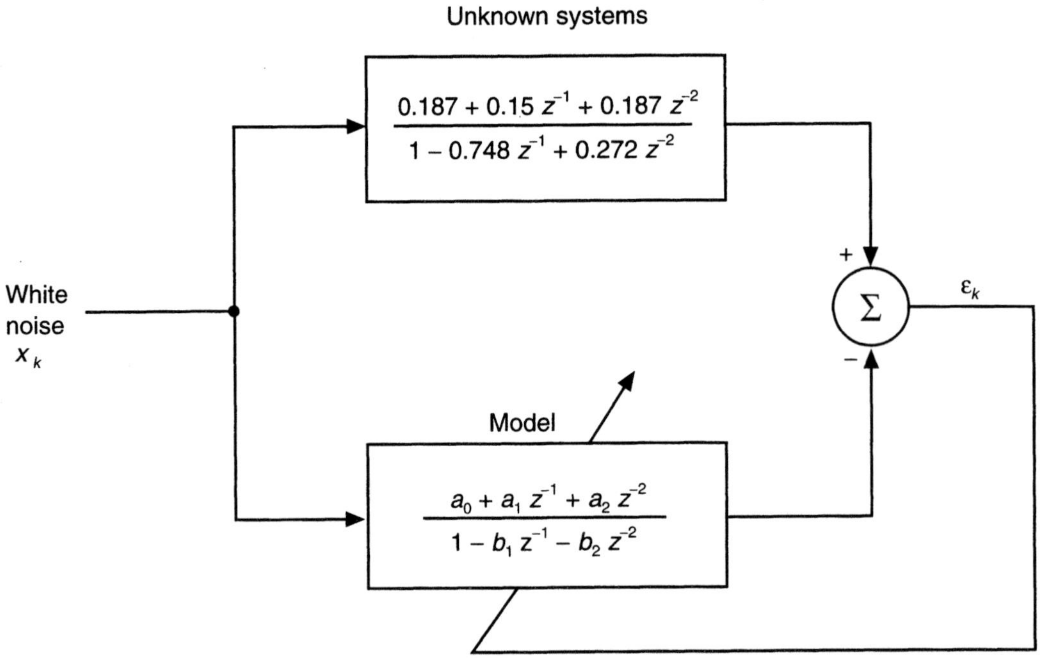

Unknown systems

$$\frac{0.187 + 0.15\,z^{-1} + 0.187\,z^{-2}}{1 - 0.748\,z^{-1} + 0.272\,z^{-2}}$$

White noise x_k

Σ + − ε_k

Model

$$\frac{a_0 + a_1\,z^{-1} + a_2\,z^{-2}}{1 - b_1\,z^{-1} - b_2\,z^{-2}}$$

FIGURE 5.3 Block diagram of adaptive system modeling example implemented by program ARMA.C.

```
#include <stdlib.h>
#include <stdio.h>
#include <math.h>
#include "rtdspc.h"

float iir_biquad(float input,float *a,float *b);
float iir_adapt_filter(float input,float d,float *a,float *b);

#define LEN 7000

void main()
{
    int i;
    float y;    static float d[LEN];
    float b[2] = { 0.7477891445, -0.2722149193 };
    float a[3] = { 0.187218, 0.149990698, 0.187218 };
```

LISTING 5.3 Program ARMA to demonstrate real-time ARMA modeling of a system. (*Continued*)

```
    /* set the random seed */
        srand(1);
        for(i = 0 ; i < LEN ; i++) d[i] = iir_biquad(gaussian(),a,b);

    /* clear all coefficients to zero at start of adaptive process */
        b[0] = b[1] = 0.0;
        a[0] = a[1] = a[2] = 0.0;

    /* reset the random seed to re-generate the same random sequence */
        srand(1);
        for(i = 0 ; i < LEN ; i++) {
            y = iir_adapt_filter(gaussian(),d[i],a,b);
            printf("\n%f %f %f %f %f %f %f",
                d[i],y,a[0],a[1],a[2],b[0],b[1]);
        }
    }
```

LISTING 5.3 *(Continued)*

```
/* 2 poles (2 b coefs) and 2 zeros (3 a coefs) iir filter single biquad */

float iir_biquad(float input,float *a,float *b)
{
    static float out_hist1,out_hist2;
    static float in_hist1,in_hist2;
    float output;

    output = out_hist1 * b[0];
    output += out_hist2 * b[1];              /* poles */

    output += input * a[0];
    output += in_hist1 * a[1];              /* zeros */
    output += in_hist2 * a[2];

    out_hist2 = out_hist1;           /* update history */
    out_hist1 = output;

    in_hist2 = in_hist1;
    in_hist1 = input;

    return(output);
}
```

LISTING 5.4 Function **iir_biquad(input,a,b)**, which implements one second-order IIR filter (contained in ARMA.C).

```
/* 2 poles (2 b coefs) and 2 zeros (3 a coefs) adaptive iir biquad filter */

float iir_adapt_filter(float input,float d,float *a,float *b)
{
    int i;
    static float out_hist1,out_hist2;
    static float beta[2],beta_h1[2],beta_h2[2];
    static float alpha[3],alpha_h1[3],alpha_h2[3];
    static float in_hist[3];
    float output,e;

    output = out_hist1 * b[0];
    output += out_hist2 * b[1];              /* poles */

    in_hist[0] = input;
    for(i = 0 ; i < 3 ; i++)
        output += in_hist[i] * a[i];          /* zeros */

/* calclulate alpha and beta update coefficients */
    for(i = 0 ; i < 3 ; i++)
        alpha[i] = in_hist[i] + b[0]*alpha_h1[i] + b[1]*alpha_h2[i];

    beta[0] = out_hist1 + b[0]*beta_h1[0] + b[1]*beta_h2[0];
    beta[1] = out_hist2 + b[0]*beta_h1[1] + b[1]*beta_h2[1];

/* error calculation */
    e = d - output;
/* update coefficients */
    a[0] += e*0.2*alpha[0];
    a[1] += e*0.1*alpha[1];
    a[2] += e*0.06*alpha[2];

    b[0] += e*0.04*beta[0];
    b[1] += e*0.02*beta[1];

/* update history for alpha */
    for(i = 0 ; i < 3 ; i++) {
        alpha_h2[i] = alpha_h1[i];
        alpha_h1[i] = alpha[i];
    }

/* update history for beta */
    for(i = 0 ; i < 2 ; i++) {
```

LISTING 5.5 Function **`iir_adapt_filter(input,d,a,b)`**, which implements an LMS adaptive second-order IIR filter (contained in ARMA.C) (*Continued*)

```
        beta_h2[i] = beta_h1[i];
        beta_h1[i] = beta[i];
    }

/* update input/output history */
    out_hist2 = out_hist1;
    out_hist1 = output;

    in_hist[2] = in_hist[1];
    in_hist[1] = input;

    return(output);
}
```

LISTING 5.5 *(Continued)*

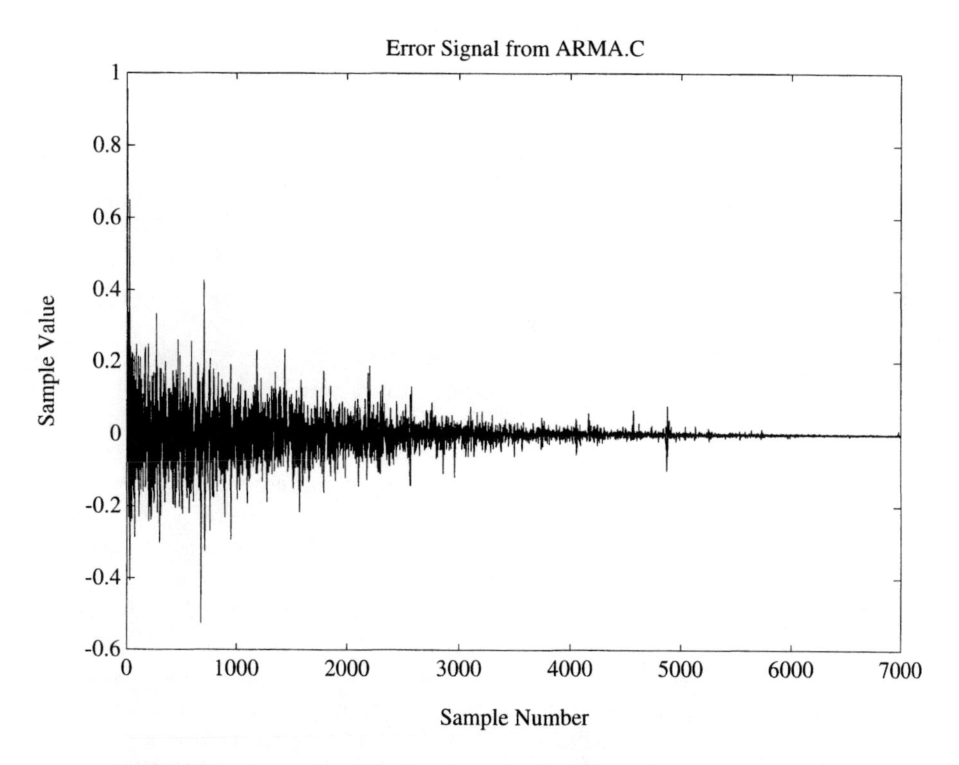

FIGURE 5.4 Error signal during the IIR adaptive process, illustrated by the program ARMA.C.

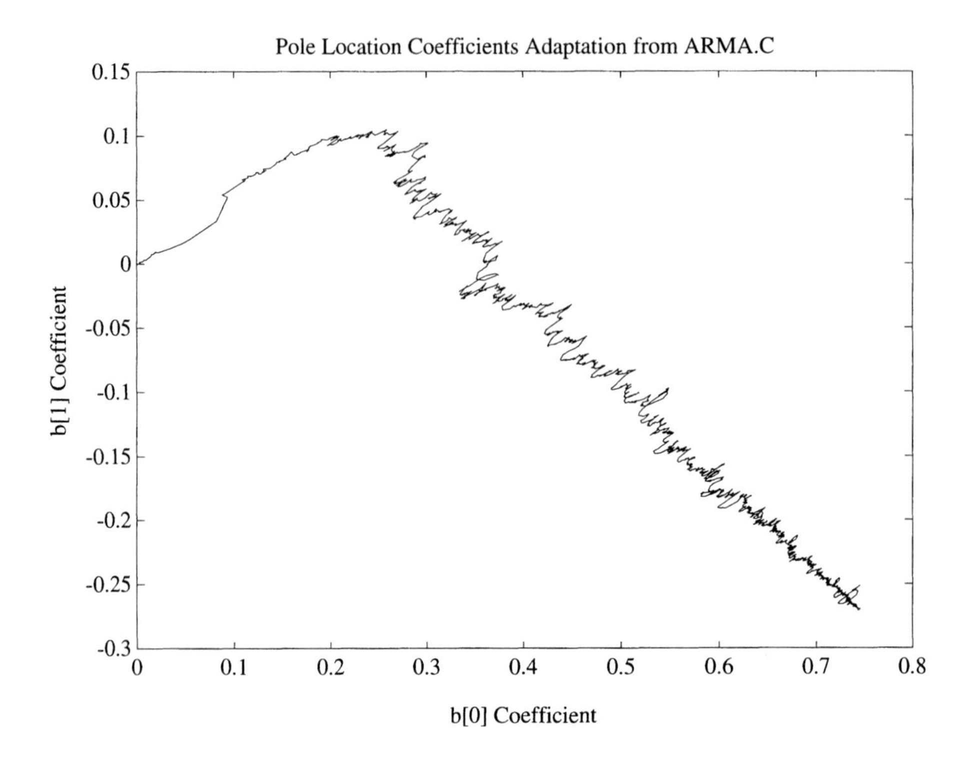

FIGURE 5.5 Pole coefficients (b0,b1) during the IIR adaptive process, illustrated by the program ARMA.C.

5.2.2 AR Frequency Estimation

The frequency of a signal can be estimated in a variety of ways using spectral analysis methods (one of which is the FFT illustrated in section 5.1.2). Another parametric approach is based on modeling the signal as resulting from an AR process with a single complex pole. The angle of the pole resulting from the model is directly related to the mean frequency estimate. This model approach can easily be biased by noise or other signals but provides a highly efficient real-time method to obtain mean frequency information.

The first step in the AR frequency estimation process is to convert the real signal input to a complex signal. This is not required when the signal is already complex, as is the case for a radar signal. Real-to-complex conversion can be done relatively simply by using a Hilbert transform FIR filter. The output of the Hilbert transform filter gives the imaginary part of the complex signal and the input signal is the real part of the complex signal. Listing 5.6 shows the program ARFREQ.C, which implements a 35-point Hilbert transform and the AR frequency estimation process. The AR frequency estimate determines the average frequency from the average phase differences between consecutive

```
#include <stdlib.h>
#include <stdio.h>
#include <string.h>
#include <math.h>
#include "rtdspc.h"

/* ARFREQ.C - take real data in one record and determine the
1st order AR frequency estimate versus time.  Uses a Hilbert transform
to convert the real signal to complex representation */

main()
{
/* 35 point Hilbert transform FIR filter cutoff at 0.02 and 0.48
   +/- 0.5 dB ripple in passband, zeros at 0 and 0.5 */

    static float  fir_hilbert35[35] = {
    0.038135,    0.000000,    0.024179,    0.000000,    0.032403,
    0.000000,    0.043301,    0.000000,    0.058420,    0.000000,
    0.081119,    0.000000,    0.120167,    0.000000,    0.207859,
    0.000000,    0.635163,    0.000000,   -0.635163,    0.000000,
   -0.207859,    0.000000,   -0.120167,    0.000000,   -0.081119,
    0.000000,   -0.058420,    0.000000,   -0.043301,    0.000000,
   -0.032403,    0.000000,   -0.024179,    0.000000,   -0.038135
                          };

    static float hist[34];
    int i,winlen;
    float sig_real,sig_imag,last_real,last_imag;
    float cpi,xr,xi,freq;

    cpi = 1.0/(2.0*PI);
    winlen = 32;

    last_real = 0.0;
    last_imag = 0.0;
    for(;;) {
/* determine the phase difference between sucessive samples */
      xr = 0.0;
      xi = 0.0;
      for(i = 0 ; i < winlen ; i++) {
        sig_imag = fir_filter(getinput(),fir_hilbert35,35,hist);
        sig_real = hist[16];
        xr += sig_real * last_real;
        xr += sig_imag * last_imag;
        xi += sig_real * last_imag;
```

LISTING 5.6 Program ARFREQ.C, which calculates AR frequency estimates in real-time. *(Continued)*

```
            xi -= sig_imag * last_real;
            last_real = sig_real;
            last_imag = sig_imag;
        }
/* make sure the result is valid, give 0 if not */
        if(fabs(xr) > 1e-10)
            freq = cpi*atan2(xi,xr);
        else
            freq = 0.0;
        sendout(freq);
    }
}
```

LISTING 5.6 *(Continued)*

complex samples. The arc tangent is used to determine the phase angle of the complex results. Because the calculation of the arc tangent is relatively slow, several simplifications can be made so that only one arc tangent is calculated for each frequency estimate. Let x_n be the complex sequence after the Hilbert transform. The phase difference is

$$\Phi_n = \arg[x_n] - \arg[x_{n-1}] = \arg[x_n x_{n-1}^*]. \tag{5.1}$$

The average frequency estimate is then

$$\hat{f} = \frac{\sum_{n=0}^{wlen-1} \Phi_n}{2\pi\, wlen} \cong \frac{\arg\left[\sum_{n=0}^{wlen-1} x_n x_{n-1}^*\right]}{2\pi}, \tag{5.2}$$

where the last approximation weights the phase differences based on the amplitude of the complex signal and reduces the number of arc tangents to one per estimate. The constant *wlen* is the window length (**`winlen`** in program ARFREQ) and controls the number of phase estimates averaged together. Figure 5.6 shows the results from the ARFREQ program when the CHKL.TXT speech data is used as input. Note that the higher frequency content of the "chi" sound is easy to identify.

5.3 SPEECH PROCESSING

Communication channels never seem to have enough bandwidth to carry the desired speech signals from one location to another for a reasonable cost. Speech compression attempts to improve this situation by sending the speech signal with as few bits per second as possible. The same channel can now be used to send a larger number of speech signals at a lower cost. Speech compression techniques can also be used to reduce the amount of memory needed to store digitized speech.

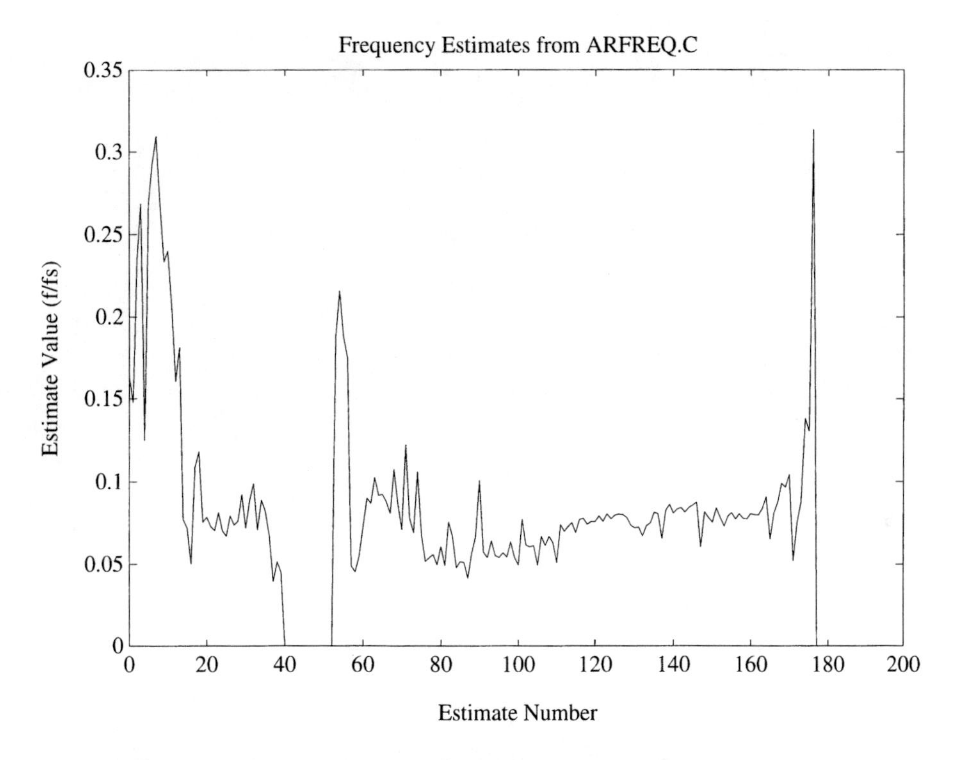

FIGURE 5.6 Frequency estimates from program ARFREQ.C, using the CHKL.DAT speech data as input.

5.3.1 Speech Compression

The simplest way to reduce the bandwidth required to transmit speech is to simply reduce the number bits per sample that are sent. If this is done in a linear fashion, then the quality of the speech (in terms of signal-to-noise ratio) will degrade rapidly when less than 8 bits per sample are used. Speech signals require 13 or 14 bits with linear quantization in order to produce a digital representation of the full range of speech signals encountered in telephone applications. The International Telegraph and Telephone Consultative Committee (CCITT, 1988) recommendation G.711 specifies the basic pulse code modulation (PCM) algorithm, which uses a logarithmic compression curve called μ-law. μ-law (see section 1.5.1 in chapter 1) is a piecewise linear approximation of a logarithmic transfer curve consisting of 8 linear segments. It compresses a 14-bit linear speech sample down to 8 bits. The sampling rate is 8000 Hz of the coded output. A compression ratio of 1.75:1 is achieved by this method without much computational complexity. Speech quality is not degraded significantly, but music and other audio signals would be degraded. Listing 5.7 shows the program MULAW.C, which encodes and decodes a speech signal using μ-law compression. The encode and decode algorithms that use tables to implement the com-

```
#include <stdlib.h>
#include <stdio.h>
#include "rtdspc.h"
#include "mu.h"

/************************************************************************

MULAW.C - PROGRAM TO DEMONSTRATE MU LAW SPEECH COMPRESSION

*************************************************************************/

main()
{
    int i,j;
    for(;;) {
        i = (int) getinput();

/* encode 14 bit linear input to mu-law */
        j = abs(i);
        if(j > 0x1fff) j = 0x1fff;
        j = invmutab[j/2];
        if(i < 0) j |= 0x80;

/* decode the 8 bit mu-law and send out */
        sendout((float)mutab[j]);
    }
}
```

LISTING 5.7 Program MULAW.C, which encodes and decodes a speech signal using μ-law compression.

pression are also shown in this listing. Because the tables are rather long, they are in the include file MU.H.

5.3.2 ADPCM (G.722)

The CCITT recommendation G.722 is a standard for digital encoding of speech and audio signals in the frequency range from 50 Hz to 7000 Hz. The G.722 algorithm uses sub-band adaptive differential pulse code modulation (ADPCM) to compress 14-bit, 16 kHz samples for transmission or storage at 64 kbits/sec (a compression ratio of 3.5:1). Because the G.722 method is a wideband standard, high-quality telephone network applications as well as music applications are possible. If the sampling rate is increased, the same algorithm can be used for good quality music compression.

The G.722 program is organized as a set of functions to optimize memory usage and make it easy to follow. This program structure is especially efficient for G.722, since most of the functions are shared between the higher and lower sub-bands. Many of the functions are also shared by both the encoder and decoder of both sub-bands. All of the functions are performed using fixed-point arithmetic, because this is specified in the CCITT recommendation. A floating-point version of the G.722 C code is included on the enclosed disk. The floating-point version runs faster on the DSP32C processor, which has limited support for the shift operator used extensively in the fixed-point implementation. Listing 5.8 shows the main program G722MAIN.C, which demonstrates the algorithm by encoding and decoding the stored speech signal "chicken little," and then operates on the real-time speech signal from **getinput()**. The output decoded signal is played using **sendout()** with an effective sample rate of 16 kHz (one sample is interpolated using simple linear interpolation giving an actual sample rate for this example of

```
#include <stdlib.h>
#include "rtdspc.h"

/* Main program for g722 encode and decode demo for 210X0 */

    extern int encode(int,int);
    extern void decode(int);
    extern void reset();

/* outputs of the decode function */
    extern int xout1,xout2;

    int chkl_coded[6000];
    extern int pm chkl[];

void main()
{
    int i,j,t1,t2;
    float xf1 = 0.0;
    float xf2 = 0.0;

/* reset, initialize required memory */
    reset();

/* code the speech, interpolate because it was recorded at 8000 Hz */
    for(i = 0 ; i < 6000 ; i++) {
        t1=64*chkl[i];
        t2=32*(chkl[i]+chkl[i+1]);
        chkl_coded[i]=encode(t1,t2);
    }

/* interpolate output to 32 KHz */
    for(i = 0 ; i < 6000 ; i++) {
```

LISTING 5.8 The main program (G722MAIN.C), which demonstrates the ADPCM algorithm in real-time. (*Continued*)

```
            decode(chkl_coded[i]);
            xf1 = (float)xout1;
            sendout(0.5*xf2+0.5*xf1);
            sendout(xf1);
            xf2 = (float)xout2;
            sendout(0.5*xf2+0.5*xf1);
            sendout(xf2);
        }

    /* simulate a 16 KHz sampling rate (actual is 32 KHz) */
    /* note: the g722 standard calls for 16 KHz for voice operation */
        while(1) {
            t1=0.5*(getinput()+getinput());
            t2=0.5*(getinput()+getinput());

            j=encode(t1,t2);
            decode(j);

            xf1 = (float)xout1;
            sendout(0.5*(xf1+xf2));
            sendout(xf1);
            xf2 = (float)xout2;
            sendout(0.5*(xf2+xf1));
            sendout(xf2);
        }
    }
```

LISTING 5.8 *(Continued)*

32 kHz). Listing 5.9 shows the **encode** function, and Listing 5.10 shows the **decode** function; both are contained in G.722.C. Listing 5.11 shows the functions **filtez**, **filtep**, **quantl**, **invqxl**, **logscl**, **scalel**, **upzero**, **uppol2**, **uppol1**, **invqah**, and **logsch**, which are used by the **encode** and **decode** functions. In Listings 5.9, 5.10, and 5.11, the global variable definitions and data tables have been omitted for clarity.

(text continues on page 215)

```
/* G722 encode function two ints in, one int out */

int encode(int xin1,int xin2)
{
    int i;
    int *h_ptr;
    int *tqmf_ptr,*tqmf_ptr1;
    long int xa,xb;
    int xl,xh;
```

LISTING 5.9 Function **encode(xin1,xin2)** (contained in G.722.C). *(Continued)*

```
    int decis;
    int         sh;          /* this comes from adaptive predictor */
    int         eh;
    int         dh;
    int         il,ih;
    int szh,sph,ph,yh;
    int szl,spl,sl,el;

/* encode: put input samples in xin1 = first value, xin2 = second value */
/* returns il and ih stored together */

/* transmit quadrature mirror filters implemented here */
    h_ptr = h;
    tqmf_ptr = tqmf;
    xa = (long)(*tqmf_ptr++) * (*h_ptr++);
    xb = (long)(*tqmf_ptr++) * (*h_ptr++);
/* main multiply accumulate loop for samples and coefficients */
    for(i = 0 ; i < 10 ; i++) {
        xa += (long)(*tqmf_ptr++) * (*h_ptr++);
        xb += (long)(*tqmf_ptr++) * (*h_ptr++);
    }
/* final mult/accumulate */
    xa += (long)(*tqmf_ptr++) * (*h_ptr++);
    xb += (long)(*tqmf_ptr) * (*h_ptr++);

/* update delay line tqmf */
    tqmf_ptr1 = tqmf_ptr - 2;
    for(i = 0 ; i < 22 ; i++) *tqmf_ptr– = *tqmf_ptr1–;
    *tqmf_ptr– = xin1;
    *tqmf_ptr = xin2;

    xl = (xa + xb) >> 15;
    xh = (xa - xb) >> 15;

/* end of quadrature mirror filter code */

/* into regular encoder segment here */
/* starting with lower sub band encoder */

/* filtez - compute predictor output section - zero section */

    szl = filtez(delay_bpl,delay_dltx);

/* filtep - compute predictor output signal (pole section) */

    spl = filtep(rlt1,al1,rlt2,al2);
```

LISTING 5.9 *(Continued)*

```
/* compute the predictor output value in the lower sub_band encoder */

    sl = szl + spl;
    el = xl - sl;

/* quantl: quantize the difference signal */

    il = quantl(el,detl);

/* invqxl: does both invqal and invqbl- computes quantized difference signal */
/* for invqbl, truncate by 2 lsbs, so mode = 3 */

/* invqal case with mode = 3 */
    dlt = ((long)detl*qq4_code4_table[il >> 2]) >> 15;

/* logscl: updates logarithmic quant. scale factor in low sub band*/
    nbl = logscl(il,nbl);

/* scalel: compute the quantizer scale factor in the lower sub band*/
/* calling parameters nbl and 8 (constant such that scalel can be scaleh) */
    detl = scalel(nbl,8);

/* parrec - simple addition to compute recontructed signal for adaptive pred */
    plt = dlt + szl;

/* upzero: update zero section predictor coefficients (sixth order)*/
/* calling parameters: dlt, dlti(circ pointer for delaying */
/* dlt1, dlt2, ..., dlt6 from dlt */
/*  bpli (linear_buffer in which all six values are delayed */
/* return params:      updated bpli, delayed dltx */

    upzero(dlt,delay_dltx,delay_bpl);

/* uppol2- update second predictor coefficient apl2 and delay it as al2 */
/* calling parameters: al1, al2, plt, plt1, plt2 */

    al2 = uppol2(al1,al2,plt,plt1,plt2);

/* uppol1 :update first predictor coefficient apl1 and delay it as al1 */
/* calling parameters: al1, apl2, plt, plt1 */

    al1 = uppol1(al1,al2,plt,plt1);

/* recons : compute recontructed signal for adaptive predictor */
    rlt = sl + dlt;

/* done with lower sub_band encoder; now implement delays for next time*/
```

LISTING 5.9 *(Continued)*

```
    rlt2 = rlt1;
    rlt1 = rlt;
    plt2 = plt1;
    plt1 = plt;

/* high band encode */

    szh = filtez(delay_bph,delay_dhx);

    sph = filtep(rh1,ah1,rh2,ah2);

/* predic: sh = sph + szh */
    sh = sph + szh;
/* subtra: eh = xh - sh */
    eh = xh - sh;

/* quanth - quantization of difference signal for higher sub-band */
/* quanth: in-place for speed params: eh, deth (has init. value) */
/* return: ih */
    if(eh >= 0) {
        ih = 3;     /* 2,3 are pos codes */
    }
    else {
        ih = 1;     /* 0,1 are neg codes */
    }
    decis = (564L*(long)deth) >> 12L;
    if(abs(eh) > decis) ih–;     /* mih = 2 case */

/* invqah: in-place compute the quantized difference signal
 in the higher sub-band*/

    dh = ((long)deth*qq2_code2_table[ih]) >> 15L ;

/* logsch: update logarithmic quantizer scale factor in hi sub-band*/

    nbh = logsch(ih,nbh);

/* note : scalel and scaleh use same code, different parameters */
    deth = scalel(nbh,10);

/* parrec - add pole predictor output to quantized diff. signal(in place)*/
    ph = dh + szh;

/* upzero: update zero section predictor coefficients (sixth order) */
/* calling parameters: dh, dhi(circ), bphi (circ) */
/* return params: updated bphi, delayed dhx */

    upzero(dh,delay_dhx,delay_bph);
```

LISTING 5.9 *(Continued)*

```
/* uppol2: update second predictor coef aph2 and delay as ah2 */
/* calling params: ah1, ah2, ph, ph1, ph2 */
/* return params:  aph2 */

    ah2 = uppol2(ah1,ah2,ph,ph1,ph2);

/* uppol1:  update first predictor coef. aph2 and delay it as ah1 */

    ah1 = uppol1(ah1,ah2,ph,ph1);

/* recons for higher sub-band */
    yh = sh + dh;

/* done with higher sub-band encoder, now Delay for next time */
    rh2 = rh1;
    rh1 = yh;
    ph2 = ph1;
    ph1 = ph;

/* multiplexing ih and il to get signals together */
    return(il | (ih << 6));
}
```

LISTING 5.9 (*Continued*)

```
/* decode function, result in xout1 and xout2 */

void decode(int input)
{
    int i;
    int xa1,xa2;     /* qmf accumulators */
    int *h_ptr;
    int pm *ac_ptr,*ac_ptr1,*ad_ptr,*ad_ptr1;
    int          ilr,ih;
    int xs,xd;
    int          rl,rh;
    int      dl;

/* split transmitted word from input into ilr and ih */
    ilr = input & 0x3f;
    ih = input >> 6;

/* LOWER SUB_BAND DECODER */

/* filtez: compute predictor output for zero section */
```

LISTING 5.10 Function **decode(input)** (contained in G.722.C). (*Continued*)

```
    dec_szl = filtez(dec_del_bpl,dec_del_dltx);

/* filtep: compute predictor output signal for pole section */

    dec_spl = filtep(dec_rlt1,dec_al1,dec_rlt2,dec_al2);

    dec_sl = dec_spl + dec_szl;

/* invqxl: compute quantized difference signal for adaptive predic in low sb */
    dec_dlt = ((long)dec_detl*qq4_code4_table[ilr >> 2]) >> 15;

/* invqxl: compute quantized difference signal for decoder output in low sb */
    dl = ((long)dec_detl*qq6_code6_table[ilr]) >> 15;

    rl = dl + dec_sl;

/* logscl: quantizer scale factor adaptation in the lower sub-band */

    dec_nbl = logscl(ilr,dec_nbl);

/* scalel: computes quantizer scale factor in the lower sub band */

    dec_detl = scalel(dec_nbl,8);

/* parrec - add pole predictor output to quantized diff. signal(in place) */
/* for partially reconstructed signal */

    dec_plt = dec_dlt + dec_szl;

/* upzero: update zero section predictor coefficients */

    upzero(dec_dlt,dec_del_dltx,dec_del_bpl);

/* uppol2: update second predictor coefficient apl2 and delay it as al2 */

    dec_al2 = uppol2(dec_al1,dec_al2,dec_plt,dec_plt1,dec_plt2);

/* uppol1: update first predictor coef. (pole setion) */

    dec_al1 = uppol1(dec_al1,dec_al2,dec_plt,dec_plt1);

/* recons : compute recontructed signal for adaptive predictor */
    dec_rlt = dec_sl + dec_dlt;

/* done with lower sub band decoder, implement delays for next time */
```

LISTING 5.10 *(Continued)*

```
    dec_rlt2 = dec_rlt1;
    dec_rlt1 = dec_rlt;
    dec_plt2 = dec_plt1;
    dec_plt1 = dec_plt;

/* HIGH SUB-BAND DECODER */

/* filtez: compute predictor output for zero section */

    dec_szh = filtez(dec_del_bph,dec_del_dhx);

/* filtep: compute predictor output signal for pole section */

    dec_sph = filtep(dec_rh1,dec_ah1,dec_rh2,dec_ah2);

/* predic:compute the predictor output value in the higher sub_band decoder */

    dec_sh = dec_sph + dec_szh;

/* invqah: in-place compute the quantized difference signal
 in the higher sub_band */

    dec_dh = ((long)dec_deth*qq2_code2_table[ih]) >> 15L ;

/* logsch: update logarithmic quantizer scale factor in hi sub band */

    dec_nbh = logsch(ih,dec_nbh);

/* scalel: compute the quantizer scale factor in the higher sub band */

    dec_deth = scalel(dec_nbh,10);

/* parrec: compute partially recontructed signal */
    dec_ph = dec_dh + dec_szh;

/* upzero: update zero section predictor coefficients */

    upzero(dec_dh,dec_del_dhx,dec_del_bph);

/*uppol2: update second predictor coefficient aph2 and delay it as ah2 */

    dec_ah2 = uppol2(dec_ah1,dec_ah2,dec_ph,dec_ph1,dec_ph2);

/* uppol1: update first predictor coef. (pole setion) */

    dec_ah1 = uppol1(dec_ah1,dec_ah2,dec_ph,dec_ph1);
```

LISTING 5.10 *(Continued)*

```
/* recons : compute recontructed signal for adaptive predictor */
    rh = dec_sh + dec_dh;

/* done with high band decode, implementing delays for next time here */
    dec_rh2 = dec_rh1;
    dec_rh1 = rh;
    dec_ph2 = dec_ph1;
    dec_ph1 = dec_ph;

/* end of higher sub_band decoder */

/* end with receive quadrature mirror filters */
    xd = rl - rh;
    xs = rl + rh;

/* receive quadrature mirror filters implemented here */
    h_ptr = h;
    ac_ptr = accumc;
    ad_ptr = accumd;
    xa1 = (long)xd * (*h_ptr++);
    xa2 = (long)xs * (*h_ptr++);
/* main multiply accumulate loop for samples and coefficients */
    for(i = 0 ; i < 10 ; i++) {
        xa1 += (long)(*ac_ptr++) * (*h_ptr++);
        xa2 += (long)(*ad_ptr++) * (*h_ptr++);
    }
/* final mult/accumulate */
    xa1 += (long)(*ac_ptr) * (*h_ptr++);
    xa2 += (long)(*ad_ptr) * (*h_ptr++);

/* scale by 2^14 */
    xout1 = xa1 >> 14;
    xout2 = xa2 >> 14;

/* update delay lines */
    ac_ptr1 = ac_ptr - 1;
    ad_ptr1 = ad_ptr - 1;
    for(i = 0 ; i < 10 ; i++) {
        *ac_ptr– = *ac_ptr1–;
        *ad_ptr– = *ad_ptr1–;
    }
    *ac_ptr = xd;
    *ad_ptr = xs;

}
```

LISTING 5.10 *(Continued)*

```
/* filtez - compute predictor output signal (zero section) */
/* input: bpl1-6 and dlt1-6, output: szl */

int filtez(int *bpl,int *dlt)
{
    int i;
    long int zl;
    zl = (long)(*bpl++) * (*dlt++);
    for(i = 1 ; i < 6 ; i++)
        zl += (long)(*bpl++) * (*dlt++);

    return((int)(zl >> 14));    /* x2 here */
}

/* filtep - compute predictor output signal (pole section) */
/* input rlt1-2 and al1-2, output spl */

int filtep(int rlt1,int al1,int rlt2,int al2)
{
    long int pl;
    pl = (long)al1*rlt1;
    pl += (long)al2*rlt2;
    return((int)(pl >> 14));    /* x2 here */
}

/* quantl - quantize the difference signal in the lower sub-band */
int quantl(int el,int detl)
{
    int ril,mil;
    long int wd,decis;

/* abs of difference signal */
    wd = abs(el);
/* determine mil based on decision levels and detl gain */
    for(mil = 0 ; mil < 30 ; mil++) {
        decis = (decis_levl[mil]*(long)detl) >> 15L;
        if(wd < decis) break;
    }
/* if mil=30 then wd is less than all decision levels */
    if(el >= 0) ril = quant26bt_pos[mil];
    else ril = quant26bt_neg[mil];
    return(ril);
}
```

LISTING 5.11 Functions used by the encode and decode algorithms of G.722 (contained in G.722.C). (*Continued*)

```
/* logscl - update the logarithmic quantizer scale factor in lower sub-band */
/* note that nbl is passed and returned */

int logscl(int il,int nbl)
{
    long int wd;
    wd = ((long)nbl * 127L) >> 7L;    /* leak factor 127/128 */
    nbl = (int)wd + wl_code_table[il >> 2];
    if(nbl < 0) nbl = 0;
    if(nbl > 18432) nbl = 18432;
    return(nbl);
}

/* scalel: compute the quantizer scale factor in the lower or upper sub-band*/

int scalel(int nbl,int shift_constant)
{
    int wd1,wd2,wd3;
    wd1 = (nbl >> 6) & 31;
    wd2 = nbl >> 11;
    wd3 = ilb_table[wd1] >> (shift_constant + 1 - wd2);
    return(wd3 << 3);
}

/* upzero - inputs: dlt, dlti[0-5], bli[0-5], outputs: updated bli[0-5] */
/* also implements delay of bli and update of dlti from dlt */

void upzero(int dlt,int *dlti,int *bli)
{
    int i,wd2,wd3;
/*if dlt is zero, then no sum into bli */
    if(dlt == 0) {
        for(i = 0 ; i < 6 ; i++) {
            bli[i] = (int)((255L*bli[i]) >> 8L); /* leak factor of 255/256 */
        }
    }
    else {
        for(i = 0 ; i < 6 ; i++) {
            if((long)dlt*dlti[i] >= 0) wd2 = 128; else wd2 = -128;
            wd3 = (int)((255L*bli[i]) >> 8L);     /* leak factor of 255/256 */
            bli[i] = wd2 + wd3;
        }
    }
/* implement delay line for dlt */
```

LISTING 5.11 *(Continued)*

```
    dlti[5] = dlti[4];
    dlti[4] = dlti[3];
    dlti[3] = dlti[2];
    dlti[2] = dlti[1];
    dlti[1] = dlti[0];
    dlti[0] = dlt;
}

/* uppol2 - update second predictor coefficient (pole section) */
/* inputs: al1, al2, plt, plt1, plt2. outputs: apl2 */

int uppol2(int al1,int al2,int plt,int plt1,int plt2)
{
    long int wd2,wd4;
    int apl2;
    wd2 = 4L*(long)al1;
    if((long)plt*plt1 >= 0L) wd2 = -wd2;    /* check same sign */
    wd2 = wd2 >> 7;                         /* gain of 1/128 */
    if((long)plt*plt2 >= 0L) {
        wd4 = wd2 + 128;                    /* same sign case */
    }
    else {
        wd4 = wd2 - 128;
    }
    apl2 = wd4 + (127L*(long)al2 >> 7L);  /* leak factor of 127/128 */

/* apl2 is limited to +-.75 */
    if(apl2 > 12288) apl2 = 12288;
    if(apl2 < -12288) apl2 = -12288;
    return(apl2);
}

/* uppol1 - update first predictor coefficient (pole section) */
/* inputs: al1, apl2, plt, plt1. outputs: apl1 */

int uppol1(int al1,int apl2,int plt,int plt1)
{
    long int wd2;
    int wd3,apl1;
    wd2 = ((long)al1*255L) >> 8L;     /* leak factor of 255/256 */
    if((long)plt*plt1 >= 0L) {
        apl1 = (int)wd2 + 192;        /* same sign case */
    }
    else {
        apl1 = (int)wd2 - 192;
    }
```

LISTING 5.11 *(Continued)*

```
/* note: wd3= .9375-.75 is always positive */
    wd3 = 15360 - apl2;               /* limit value */
    if(apl1 > wd3) apl1 = wd3;
    if(apl1 < -wd3) apl1 = -wd3;
    return(apl1);
}

/* logsch - update the logarithmic quantizer scale factor in higher sub-band */
/* note that nbh is passed and returned */

int logsch(int ih,int nbh)
{
    int wd;
    wd = ((long)nbh * 127L) >> 7L;        /* leak factor 127/128 */
    nbh = wd + wh_code_table[ih];
    if(nbh < 0) nbh = 0;
    if(nbh > 22528) nbh = 22528;
    return(nbh);
}
```

LISTING 5.11 *(Continued)*

Figure 5.7 shows a block diagram of the G.722 encoder (transmitter), and Figure 5.8 shows a block diagram of the G.722 decoder (receiver). The entire algorithm has six main functional blocks, many of which use the same functions:

(1) A transmit quadrature mirror filter (QMF) that splits the frequency band into two sub-bands.

(2&3) A lower sub-band encoder and higher sub-band encoder that operate on the data produced by the transmit QMF.

(4&5) A lower sub-band decoder and higher sub-band decoder.

(6) A receive QMF that combines the outputs of the decoder into one value.

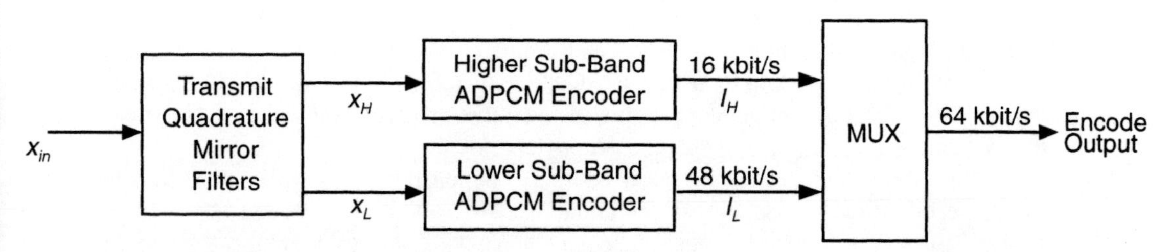

FIGURE 5.7 Block diagram of ADPCM encoder (transmitter) implemented by program G.722.C.

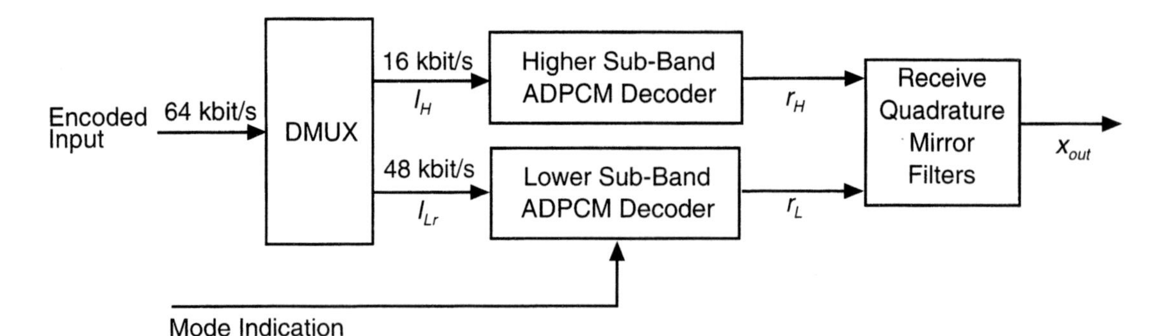

FIGURE 5.8 Block diagram of ADPCM decoder (receiver) implemented by program G.722.C.

The G.722.C functions have been checked against the G.722 specification and are fully compatible with the CCITT recommendation. The functions and program variables are named according to the functional blocks of the algorithm specification whenever possible.

Quadrature mirror filters are used in the G.722 algorithm as a method of splitting the frequency band into two sub-bands (higher and lower). The QMFs also decimate the encoder input from 16 kHz to 8 kHz (*transmit QMF*) and interpolate the decoder output from 8 kHz to 16 kHz (*receive QMF*). These filters are 24-tap FIR filters whose impulse response can be considered lowpass and highpass filters. Both the transmit and receive QMFs share the same coefficients and a delay line of the same number of taps.

Figure 5.9 shows a block diagram of the higher sub-band encoder. The lower and higher sub-band encoders operate on an estimated difference signal. The number of bits required to represent the difference is smaller than the number of bits required to represent the complete input signal. This difference signal is obtained by subtracting a predicted value from the input value:

```
el = xl - sl
eh = xh - sh
```

The predicted value, **sl** or **sh,** is produced by the adaptive predictor, which contains a second-order filter section to model poles, and a sixth-order filter section to model zeros in the input signal. After the predicted value is determined and subtracted from the input signal, the estimate signal **el** is applied to a nonlinear adaptive quantizer.

One important feature of the sub-band encoders is a feedback loop. The output of the adaptive quantizer is fed to an inverse adaptive quantizer to produce a difference signal. This difference signal is then used by the adaptive predictor to produce **sl** (the estimate of the input signal) and update the adaptive predictor.

The G.722 standard specifies an auxiliary, nonencoded data channel. While the G.722 encoder always operates at an output rate of 64 kbits per second (with 14-bit, 16kHz input samples), the decoder can accept encoded signals at 64, 56, or 48 kbps. The 56 and 48 kbps bit rates correspond to the use of the auxiliary data channel, which operates at either 8 or 16 kbps. A *mode indication signal* informs the decoder which mode is being used. This feature is not implemented in the G.722.C program.

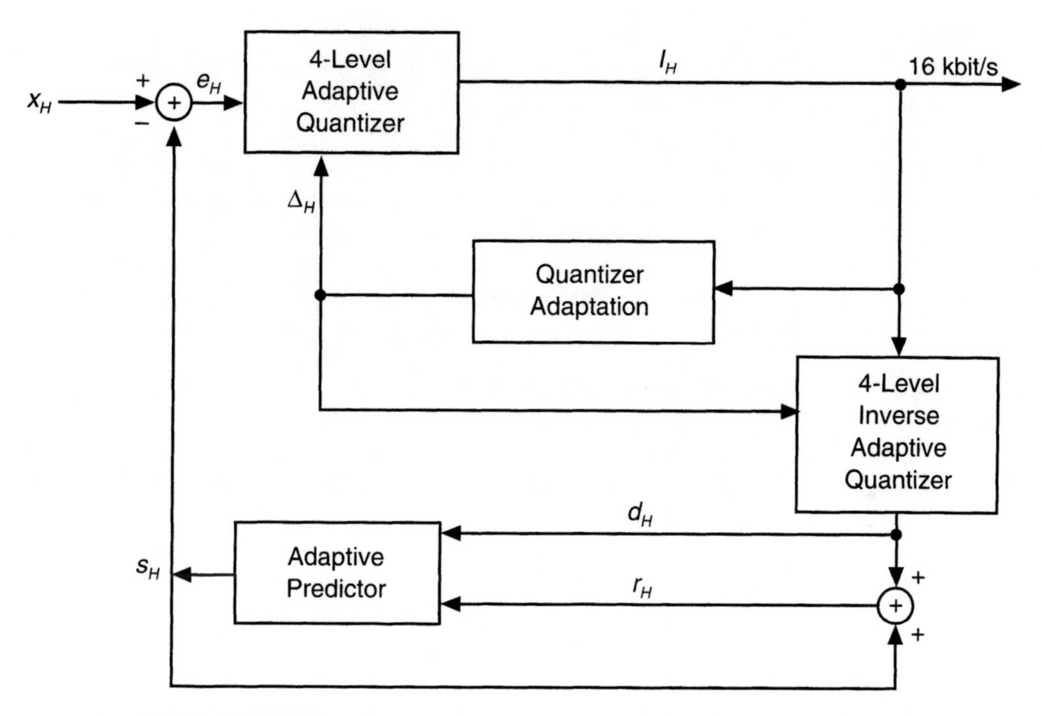

FIGURE 5.9 Block diagram of higher sub-band ADPCM encoder implemented by program G.722.C.

Figure 5.10 shows a block diagram of the higher sub-band decoder. In general, both the higher and lower sub-band encoders and decoders make the same function calls in almost the same order because they are similar in operation. For mode 1, a 60 level inverse adaptive quantizer is used in the lower sub-band, which gives the best speech quality. The higher sub-band uses a 4 level adaptive quantizer.

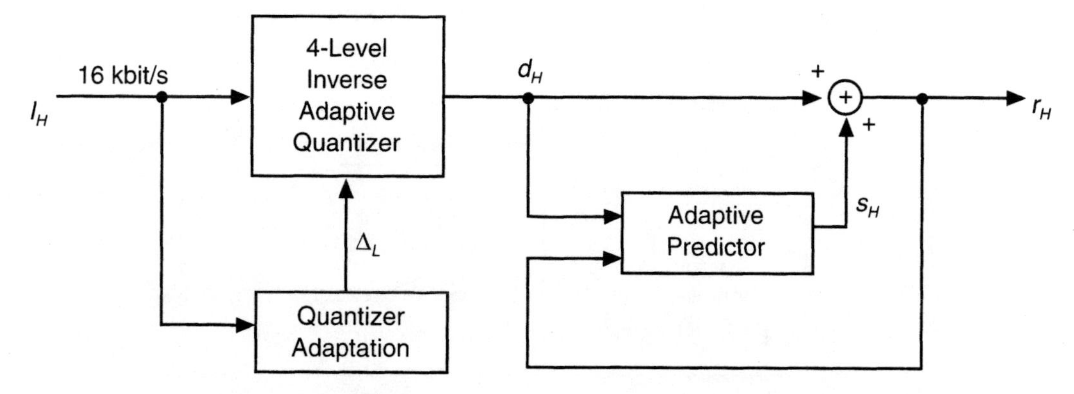

FIGURE 5.10 Block diagram of higher sub-band ADPCM decoder implemented by program G.722.C.

5.4 MUSIC PROCESSING

Music signals require more dynamic range and a much wider bandwidth than speech signals. Professional quality music processing equipment typically uses 18 to 24 bits to represent each sample and a 48 kHz or higher sampling rate. Consumer digital audio processing (in CD players, for example) is usually done with 16-bit samples and a 44.1 kHz sampling rate. In both cases, music processing is a far greater challenge to a digital signal processor than speech processing. More MIPs are required for each operation and quantization noise in filters becomes more important. In most cases DSP techniques are less expensive and can provide a higher level of performance than analog techniques.

5.4.1 Equalization and Noise Removal

Equalization refers to a filtering process where the frequency content of an audio signal is adjusted to make the source sound better, adjust for room acoustics, or remove noise that may be in a frequency band different from the desired signal. Most audio equalizers have a number of bands that operate on the audio signal in parallel with the output of each filter, added together to form the equalized signal. This structure is shown in Figure 5.11. The program EQUALIZ.C is shown in Listing 5.12. Each **gain** constant is used to adjust the relative signal amplitude of the output of each bandpass filter. The input signal is always added to the output such that if all the gain values are zero the signal is unchanged. Setting the **gain** values greater than zero will boost frequency response in each band. For example, a boost of 6 dB is obtained by setting one of thc **gain** values to 1.0. The center frequencies and number of bandpass filters in analog audio equalizers vary widely from one manufacturer to another. A seven-band equalizer with center frequencies at 60,

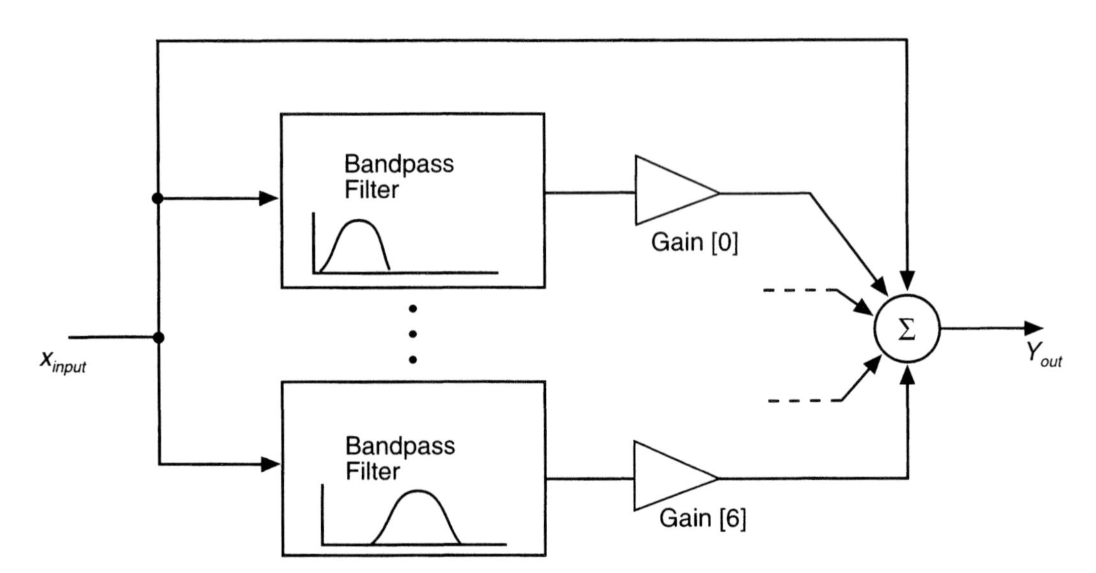

FIGURE 5.11 Block diagram of audio equalization implemented by program EQUALIZ.C.

```
#include <stdlib.h>
#include <math.h>
#include "rtdspc.h"

/**************************************************************************

EQUALIZ.C - PROGRAM TO DEMONSTRATE AUDIO EQUALIZATION
            USING 7 IIR BANDPASS FILTERS.

*************************************************************************/

/* gain values global so they can be changed in real-time */
/* start at flat pass through */
    float gain[7] = { 0.0, 0.0, 0.0, 0.0, 0.0, 0.0, 0.0 };

void main()
{
    int i;
    float signal_in,signal_out;

/* history arrays for the filters */
    static float hist[7][2];

/* bandpass filter coefficients for a 44.1 kHz sampling rate */
/* center freqs are 60, 150, 400, 1000, 2400, 6000, 15000 Hz */
/* at other rates center freqs are:                          */
/* at 32 kHz:       44, 109, 290,  726, 1742, 4354, 10884 Hz */
/* at 22.1 kHz:     30,  75, 200,  500, 1200, 3000,  7500 Hz */

    static float bpf[7][5] = {
  { 0.0025579741, -1.9948111773,  0.9948840737,  0.0,  -1.0 },
  { 0.0063700872, -1.9868060350,  0.9872598052,  0.0,  -1.0 },
  { 0.0168007612, -1.9632060528,  0.9663984776,  0.0,  -1.0 },
  { 0.0408578217, -1.8988473415,  0.9182843566,  0.0,  -1.0 },
  { 0.0914007276, -1.7119922638,  0.8171985149,  0.0,  -1.0 },
  { 0.1845672876, -1.0703823566,  0.6308654547,  0.0,  -1.0 },
  { 0.3760778010,  0.6695288420,  0.2478443682,  0.0,  -1.0 },
            };

    for(;;) {
/* sum 7 bpf outputs + input for each new sample */
        signal_out = signal_in = getinput();
        for(i = 0 ; i < 7 ; i++)
          signal_out += gain[i]*iir_filter(signal_in,bpf[i],1,hist[i]);
        sendout(signal_out);
    }
}
```

LISTING 5.12 Program EQUALIZ.C, which performs equalization on audio samples in real-time.

150, 400, 1000, 2400, 6000, and 15000 Hz is implemented by program EQUALIZ.C. The bandwidth of each filter is 60 percent of the center frequency in each case, and the sampling rate is 44100 Hz. This gives the coefficients in the example equalizer program EQUALIZ.C shown in Listing 5.12. The frequency response of the 7 filters is shown in Figure 5.12.

5.4.2 Pitch-Shifting

Changing the pitch of a recorded sound is often desired in order to allow it to mix with a new song, or for special effects where the original sound is shifted in frequency to a point where it is no longer identifiable. New sounds are often created by a series of pitch shifts and mixing processes.

Pitch-shifting can be accomplished by interpolating a signal to a new sampling rate, and then playing the new samples back at the original sampling rate (see Alles, 1980; or Smith and Gossett, 1984). If the pitch is shifted down (by an interpolation factor greater than one), the new sound will have a longer duration. If the pitch is shifted upward (by an interpolation factor less than one where some decimation is occurring), the sound becomes shorter. Listing 5.13 shows the program PSHIFT.C, which can be used to pitch-

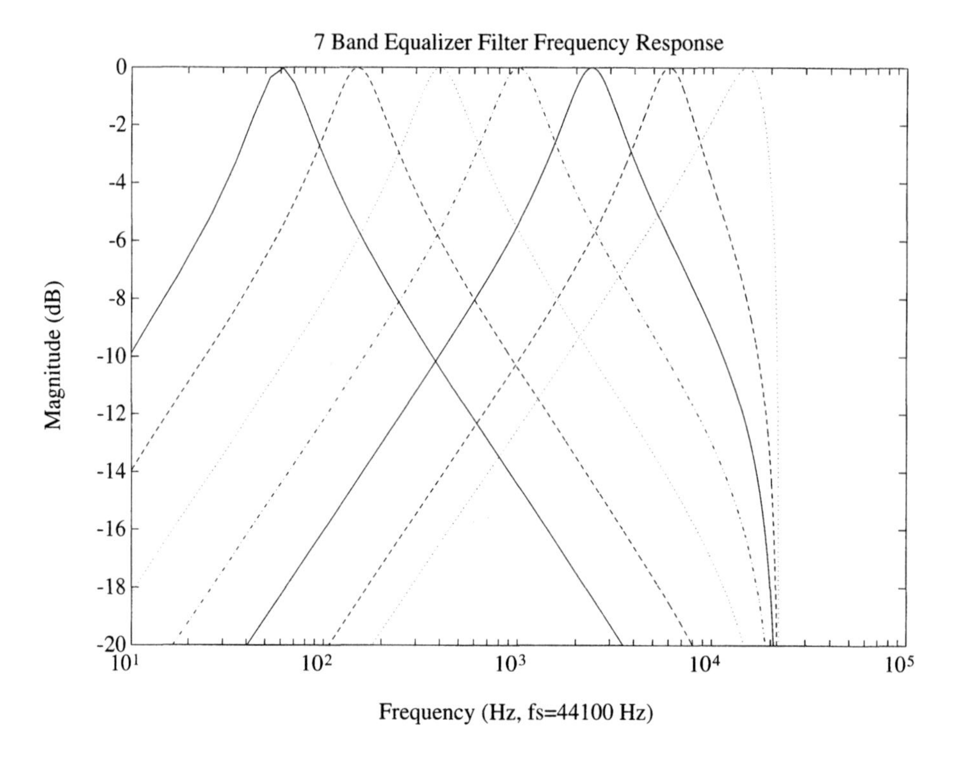

FIGURE 5.12 Frequency response of 7 filters used in program EQUALIZ.C.

shift a sound up or down by any number of semitones (12 semitones is an octave as indicated by equation 4.13). It uses a long Kaiser window filter for interpolation of the samples as illustrated in section 4.3.2 in chapter 4. The filter coefficients are calculated in the first part of the PSHIFT program before real-time input and output begins. The filtering is done with two FIR filter functions, which are shown in Listing 5.14. The history array is only updated when the interpolation point moves to the next input sample. This requires that the history update be removed from the **fir_filter** function discussed previously. The history is updated by the function **fir_history_update**. The coefficients are decimated into short polyphase filters. An interpolation ratio of up to 300 is performed and the decimation ratio is determined by the amount of pitch shift selected by the integer variable **key**.

```
#include <stdlib.h>
#include <math.h>
#include "rtdspc.h"

/* Kaiser Window Pitch Shift Algorithm */

/* set interpolation ratio */
    int ratio = 300;
/* passband specified, larger makes longer filters */
    float percent_pass = 80.0;
/* minimum attenuation in stopbands (dB), larger make long filters */
    float att = 50.0;
/* key value to shift by (semi-tones up or down) */
/* 12 is one octave */
    int key = -12;
    int lsize;

void main()
{
    int i,j;
    int nfilt,npair,n,k;
    float fa,fp,deltaf,beta,valizb,alpha;
    float w,ck,y,npair_inv,pi_ratio;
    float signal_in,phase,dec;
    int old_key = 0;            /* remember last key value */
    float **h;

    static float hist[100];     /* lsize can not get bigger than 100 */

    long int filter_length(float,float,float *);
    float izero(float);
```

LISTING 5.13 Program PSHIFT.C, which performs pitch shifting on audio samples in real-time. (*Continued*)

```
    float fir_filter_no_update(float input,float *coef,int n,float *history);
    void fir_update_history(float input,int n,float *history);

    fp = percent_pass/(200.0*ratio);
    fa = (200.0 - percent_pass)/(200.0*ratio);
    deltaf = fa-fp;

    nfilt = filter_length( att, deltaf, &beta );

    lsize = nfilt/ratio;

    nfilt = (long)lsize*ratio + 1;
    npair = (nfilt - 1)/2;

    h = (float **) calloc(ratio,sizeof(float *));
    if(!h) exit(1);
    for(i = 0 ; i < ratio ; i++) {
        h[i] = (float *) calloc(lsize,sizeof(float));
        if(!h[i]) exit(1);
    }

    /* Compute Kaiser window sample values */
    i = 0;
    j = 0;
    valizb = 1.0 / izero(beta);
    npair_inv = 1.0/npair;
    pi_ratio = PI/ratio;
    h[i++][j] = 0.0;          /* n = 0 case */
    for (n = 1 ; n < npair ; n++) {
        k = npair - n;
        alpha = k * npair_inv;
        y = beta * sqrt(1.0 - (alpha * alpha));
        w = valizb * izero(y);
        ck = ratio*sin(k*pi_ratio)/(k*PI);
        h[i++][j] = w * ck;
        if(i == ratio) {
            i = 0;
            j++;
        }
    }
/* force the pass through point */
    h[i][lsize/2] = 1.0;

/* second half of response */
    for(n = 1; n < npair; n++) {
        i = npair - n;          /* "from" location */
```

LISTING 5.13 *(Continued)*

```
        k = npair + n;              /* "to" location */
        h[k%ratio][k/ratio] = h[i%ratio][i/ratio];
    }

/* interpolate the data by calls to fir_filter_no_update,
decimate the interpolated samples by only generating the samples
required */
    phase = 0.0;
    dec = (float)ratio;
    for( ; ;) {

/* decimation ratio for key semitones shift */
/* allow real-time updates */
        if(key != old_key) {
            dec = ratio*pow(2.0,0.0833333333*key);
            old_key = key;
        }

        signal_in = getinput();
        while(phase < (float)ratio) {
            k = (int)phase;     /* pointer to poly phase values */
            sendout(fir_filter_no_update(signal_in,h[k],lsize,hist));
            phase += dec;
        }
        phase -= ratio;
        fir_update_history(signal_in,lsize,hist);
    }
}

/* Use att to get beta (for Kaiser window function) and nfilt (always odd
    valued and = 2*npair +1) using Kaiser's empirical formulas.          */
long int filter_length(float att,float deltaf,float *beta)
{
    long int npair;
    *beta = 0.0;        /* value of beta if att < 21 */
    if(att >= 50.0) *beta = .1102 * (att - 8.71);
    if (att < 50.0 & att >= 21.0)
        *beta = .5842 * pow( (att-21.0), 0.4) + .07886 * (att - 21.0);
    npair = (long int)( (att - 8.0) / (28.72 * deltaf) );
    return(2*npair + 1);
}

/* Compute Bessel function Izero(y) using a series approximation */
float izero(float y){
    float s=1.0, ds=1.0, d=0.0;
    do {
```

LISTING 5.13 *(Continued)*

```
        d = d + 2;
        ds = ds * (y*y)/(d*d);
        s = s + ds;
    } while( ds > 1E-7 * s);
    return(s);
}
```

LISTING 5.13 *(Continued)*

```
/* run the fir filter and do not update the history array */

float fir_filter_no_update(float input,float *coef,int n,float *history)
{
    int i;
    float *hist_ptr,*coef_ptr;
    float output;

    hist_ptr = history;
    coef_ptr = coef + n - 1;            /* point to last coef */

/* form output accumulation */
    output = *hist_ptr++ * (*coef_ptr--);
    for(i = 2 ; i < n ; i++) {
        output += (*hist_ptr++) * (*coef_ptr--);
    }
    output += input * (*coef_ptr);       /* input tap */

    return(output);
}

/* update the fir_filter history array */

void fir_update_history(float input,int n,float *history)
{
    int i;
    float *hist_ptr,*hist1_ptr;

    hist_ptr = history;
    hist1_ptr = hist_ptr;               /* use for history update */
    hist_ptr++;

    for(i = 2 ; i < n ; i++) {
        *hist1_ptr++ = *hist_ptr++;      /* update history array */
    }
    *hist1_ptr = input;                 /* last history */
}
```

LISTING 5.14 Functions **fir_filter_no_update** and **fir_filter_update_history** used by program PSHIFT.C.

5.4.3 Music Synthesis

Music synthesis is a natural DSP application because no input signal or A/D converter is required. Music synthesis typically requires that many different sounds be generated at the same time and mixed together to form a chord or multiple instrument sounds (see Moorer, 1977). Each different sound produced from a synthesis is referred to as a voice. The duration and starting point for each voice must be independently controlled. Listing 5.15 shows the program MUSIC.C, which plays a sound with up to 6 voices at the same time. It uses the function **note** (see Listing 5.16) to generate samples from a second order IIR oscillator using the same method as discussed in section 4.5.1 in chapter 4. The envelope of each note is specified using break points. The array **trel** gives the relative times when the amplitude should change to the values specified in array **amps**. The envelope will grow and decay to reach the amplitude values at each time specified based on the calculated first-order constants stored in the **rates** array. The frequency of the second order oscillator in **note** is specified in terms of the semitone note number **key**. A **key** value of 69 will give 440 Hz, which is the musical note A above middle C.

```
#include <stdlib.h>
#include <math.h>
#include "rtdspc.h"
#include "song.h"          /* song[108][7] array */

/* 6 Voice Music Generator */

typedef struct {
    int key,t,cindex;
    float cw,a,b;
    float y1,y0;
} NOTE_STATE;

#define MAX_VOICES 6

    float note(NOTE_STATE *,int *,float *);

void main()
{
    long int n,t,told;
    int vnum,v,key;
    float ampold;
    register long int i,endi;
    register float sig_out;
```

LISTING 5.15 Program MUSIC.C, which illustrates music synthesis by playing a 6-voice song. (*Continued*)

```
    static NOTE_STATE notes[MAX_VOICES*SONG_LENGTH];

    static float trel[5] = {    0.1 ,    0.2,    0.7,  1.0, 0.0 };
    static float amps[5] = { 3000.0 , 5000.0, 4000.0, 10.0, 0.0 };
    static float rates[10];
    static int tbreaks[10];

    for(n = 0 ; n < SONG_LENGTH ; n++) {

/* number of samples per note */
        endi = 6*song[n][0];

/* calculate the rates required to get the desired amps */
        i = 0;
        told = 0;
        ampold = 1.0;        /* always starts at unity */
        while(amps[i] > 1.0) {
            t = trel[i]*endi;
            rates[i] = exp(log(amps[i]/ampold)/(t-told));
            ampold = amps[i];
            tbreaks[i] = told = t;
            i++;
        }

/* set the key numbers for all voices to be played (vnum is how many) */
        for(v = 0 ; v < MAX_VOICES ; v++) {
            key = song[n][v+1];
            if(!key) break;
            notes[v].key = key;
            notes[v].t = 0;
        }
        vnum = v;

        for(i = 0 ; i < endi ; i++) {
            sig_out = 0.0;
            for(v = 0 ; v < vnum ; v++) {
                sig_out += note(&notes[v],tbreaks,rates);
            }
            sendout(sig_out);
        }
    }
    flush();
}
```

LISTING 5.15 *(Continued)*

```
#include <stdlib.h>
#include <math.h>
#include "rtdspc.h"

/* Function to generate samples from a second order oscillator */

/* key constant is 1/12 */
#define KEY_CONSTANT 0.083333333333333

/* this sets the A above middle C reference frequency of 440 Hz */
#define TWO_PI_DIV_FS_440  (880.0 * PI / SAMPLE_RATE)

/* cw is the cosine constant required for fast changes of envelope */
/* a and b are the coefficients for the difference equation */
/* y1 and y0 are the history values */
/* t is time index for this note */
/* cindex is the index into rate and tbreak arrays (reset when t=0) */

typedef struct {
    int key,t,cindex;
    float cw,a,b;
    float y1,y0;
} NOTE_STATE;

/*
key:
    semi-tone pitch to generate,
    number 69 will give A above middle C at 440 Hz.
rate_array:
    rate constants determines decay or rise of envelope (close to 1)
tbreak_array:
    determines time index when to change rate
*/

/* NOTE_STATE structure, time break point array, rate parameter array */

float note(NOTE_STATE *s,int *tbreak_array,float *rate_array)
{
    register int ti,ci;
    float wosc,rate,out;

    ti = s->t;
/* t=0 re-start case, set state variables */
    if(!ti) {
        wosc = TWO_PI_DIV_FS_440 * pow(2.0,(s->key-69) * KEY_CONSTANT);
```

LISTING 5.16 Function **note(state,tbreak_array,rate_array)** generates the samples for each note in the MUSIC.C program. (*Continued*)

```
        s->cw = 2.0 * cos(wosc);
        rate = rate_array[0];
        s->a = s->cw * rate;        /* rate change */
        s->b = -rate * rate;
        s->y0 = 0.0;
        out = rate*sin(wosc);
        s->cindex = 0;
    }
    else {
        ci = s->cindex;
    /* rate change case */
        if(ti == tbreak_array[ci]) {
            rate = rate_array[++ci];
            s->a = s->cw * rate;
            s->b = -rate * rate;
            s->cindex = ci;
        }

    /* make new sample */
        out = s->a * s->y1 + s->b * s->y0;
        s->y0 = s->y1;
    }
    s->y1 = out;
    s->t = ++ti;
    return(out);
}
```

LISTING 5.16 *(Continued)*

5.5 ADAPTIVE FILTER APPLICATIONS

A signal can be effectively improved or enhanced using adaptive methods, if the signal frequency content is narrow compared to the bandwidth and the frequency content changes with time. If the frequency content does not change with time, a simple matched filter will usually work better with less complexity. The basic LMS algorithm is illustrated in the next section. Section 5.5.2 illustrates a method that can be used to estimate the changing frequency of a signal using an adaptive LMS algorithm.

5.5.1 LMS Signal Enhancement

Figure 5.13 shows the block diagram of an LMS adaptive signal enhancement that will be used to illustrate the basic LMS algorithm. This algorithm was described in section 1.7.2 in chapter 1. The input signal is a sine wave with added white noise. The adaptive LMS

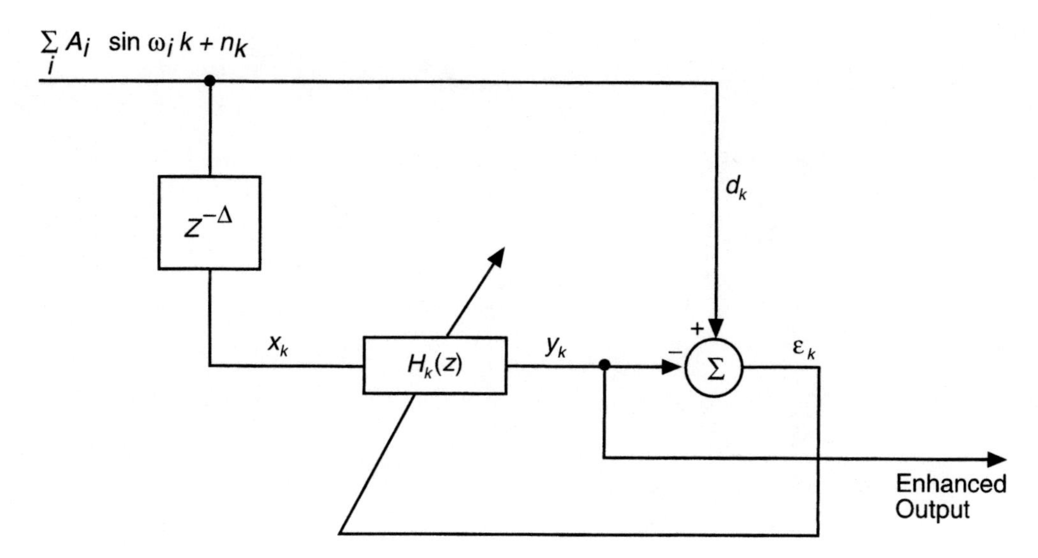

FIGURE 5.13 Block diagram of LMS adaptive signal enhancement.

```
#include <stdlib.h>
#include <stdio.h>
#include <math.h>
#include "rtdspc.h"

#define N 351
#define L 20                    /* filter order, (length L+1) */

/* set convergence parameter */
    float mu = 0.01;

void main()
{
    float lms(float,float,float *,int,float,float);
    static float d[N],b[21];
    float signal_amp,noise_amp,arg,x,y;
    int k;

/* create signal plus noise */
    signal_amp = sqrt(2.0);
    noise_amp = 0.2*sqrt(12.0);
    arg = 2.0*PI/20.0;
```

LISTING 5.17 Program LMS.C which illustrates signal-to-noise enhancement using the LMS algorithm. (*Continued*)

```
        for(k = 0 ; k < N ; k++) {
            d[k] = signal_amp*sin(arg*k) + noise_amp*gaussian();
        }

    /* scale based on L */
        mu = 2.0*mu/(L+1);

        x = 0.0;
        for(k = 0 ; k < N ; k++) {
            sendout(lms(x,d[k],b,L,mu,0.01));
    /* delay x one sample */
            x = d[k];
        }
    }
```

LISTING 5.17 *(Continued)*

```
/*
       function lms(x,d,b,l,mu,alpha)

Implements NLMS Algorithm b(k+1)=b(k)+2*mu*e*x(k)/((l+1)*sig)

x      = input data
d      = desired signal
b[0:l] = Adaptive coefficients of lth order fir filter
l      = order of filter (> 1)
mu     = Convergence parameter (0.0 to 1.0)
alpha  = Forgetting factor   sig(k)=alpha*(x(k)**2)+(1-alpha)*sig(k-1)
           (>= 0.0 and < 1.0)

returns the filter output
*/

float lms(float x,float d,float *b,int l,
                  float mu,float alpha)
{
    int ll;
    float e,mu_e,lms_const,y;
    static float px[51];       /* max L = 50 */
    static float sigma = 2.0; /* start at 2 and update internally */

    px[0]=x;
```

LISTING 5.18 Function **lms(x,d,b,l,mu,alpha)** implements the LMS algorithm. *(Continued)*

```
/* calculate filter output */
    y=b[0]*px[0];
    for(ll = 1 ; ll <= l ; ll++)
        y=y+b[ll]*px[ll];

/* error signal */
    e=d-y;

/* update sigma */
    sigma=alpha*(px[0]*px[0])+(1-alpha)*sigma;
    mu_e=mu*e/sigma;

/* update coefficients */
    for(ll = 0 ; ll <= l ; ll++)
        b[ll]=b[ll]+mu_e*px[ll];
/* update history */
    for(ll = l ; ll >= 1 ; ll—)
        px[ll]=px[ll-1];

    return(y);
}
```

LISTING 5.18 *(Continued)*

algorithm (see Listings 5.17 and 5.18) is a 21 tap (20th order) FIR filter where the filter coefficients are updated with each sample. The desired response in this case is the noisy signal and the input to the filter is a delayed version of the input signal. The delay (Δ) is selected so that the noise components of d_k and x_k are uncorrelated (a one-sample delay works well for sine waves and white noise).

The convergence parameter **mu** is the only input to the program. Although many researchers have attempted to determine the best value for **mu**, no universal solution has been found. If **mu** is too small, the system may not converge rapidly to a signal, as is illustrated in Figure 5.14. The adaptive system is moving from no signal (all coefficients are zero) to an enhanced signal. This takes approximately 300 samples in Figure 5.14b with **mu** = 0.01 and approximately 30 samples in Figure 5.14c with **mu** = 0.1.

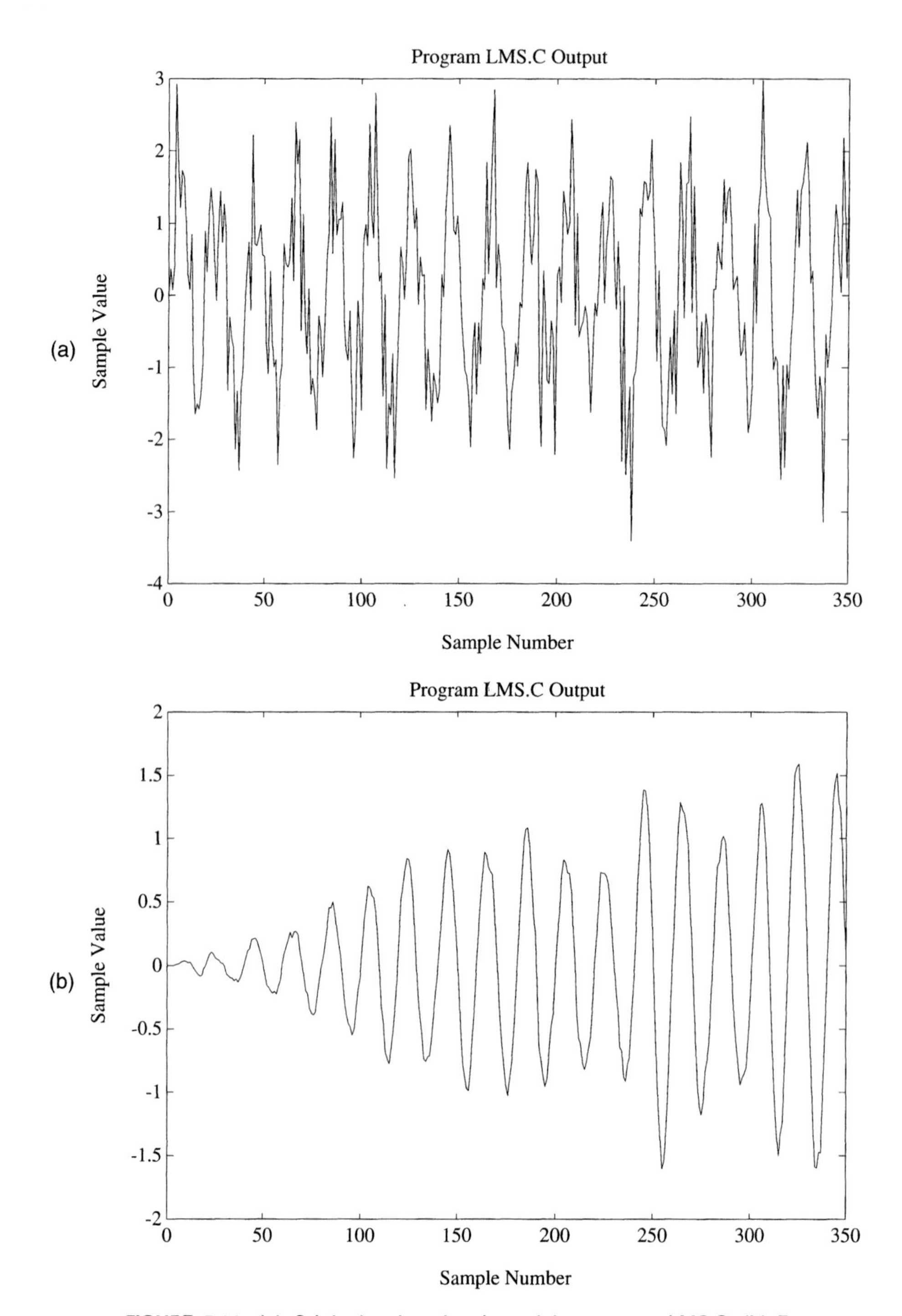

FIGURE 5.14 **(a)** Original noisy signal used in program LMS.C. **(b)** Enhanced signal obtained from program LMS.C with `mu` = 0.01.

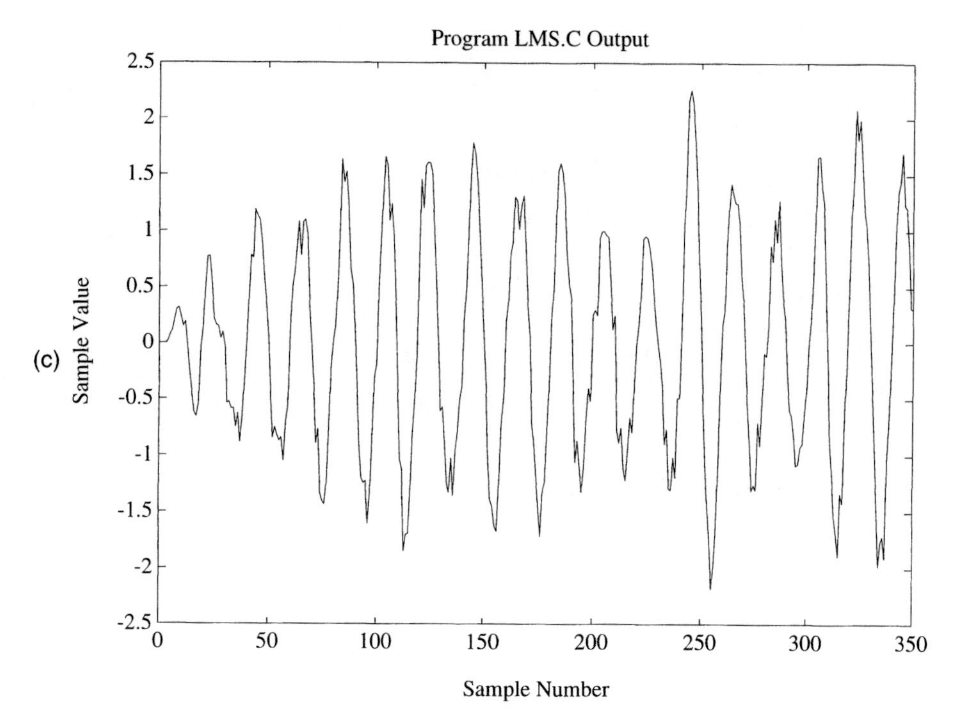

FIGURE 5.14 **(c)** Enhanced signal obtained from program LMS.C with `mu` = 0.1. *(Continued)*

5.5.2 Frequency Tracking with Noise

Listing 5.19 shows the INSTF.C program, which uses the **`lms`** function to determine instantaneous frequency estimates. Instead of using the output of the adaptive filter as illustrated in the last section, the INSTF program uses the filter coefficients to estimate the frequency content of the signal. A 1024-point FFT is used to determine the frequency response of the adaptive filter every 100 input samples. The same peak location finding algorithm as used in section 5.1.2 is used to determine the interpolated peak frequency response of the adaptive filter. Note that because the filter coefficients are real, only the first half of the FFT output is used for the peak search.

Figure 5.15 shows the output of the INSTF program when the 100,000 samples from the OSC.C program (see section 4.5.1 of chapter 4) are provided as an input. Figure 5.15(a) shows the result without added noise, and Figure 5.15(b) shows the result when white Gaussian noise (standard deviation = 100) is added to the signal from the OSC program. Listing 5.19 shows how the INSTF program was used to add the noise to the input signal using the **`gaussian()`** function. Note the positive bias in both results due to the finite length (128 in this example) of the adaptive FIR filter. Also, in Figure 5.15(b) the first few estimates are off scale because of the low signal level in the beginning portion of the waveform generated by the OSC program (the noise dominates in the first 10 estimates).

```
#include <stdlib.h>
#include <stdio.h>
#include <math.h>
#include "rtdspc.h"

/* LMS Instantaneous Frequency Estimation Program */

#define L 127        /* filter order, L+1 coefficients */
#define LMAX 200     /* max filter order, L+1 coefficients */
#define NEST 100     /* estimate decimation ratio in output */

/* FFT length must be a power of 2 */
#define FFT_LENGTH 1024
#define M 10            /* must be log2(FFT_LENGTH) */

/* set convergence parameter */
    float mu = 0.01;

void main()
{
    float lms(float,float,float *,int,float,float);
    static float b[LMAX];
    static COMPLEX samp[FFT_LENGTH];
    static float  mag[FFT_LENGTH];
    float x,d,tempflt,p1,p2;
    int i,j,k;

/* scale based on L */
    mu = 2.0*mu/(L+1);

    x = 0.0;
    for(;;) {
        for(i = 0 ; i < NEST ; i++) {
/* add noise to input for this example */
            x = getinput() + 100.0*gaussian();
            lms(x,d,b,L,mu,0.01);
/* delay d one sample */
            d = x;
        }
```

LISTING 5.19 Program INSTF.C. which uses function **lms(x,d,b,l,mu,alpha)** to implement the LMS frequency tracking algorithm. (*Continued*)

```
/* copy L+1 coefficients */
        for(i = 0 ; i <= L ; i++) {
            samp[i].real = b[i];
            samp[i].imag = 0.0;
        }

/* zero pad */
        for( ; i < FFT_LENGTH ; i++) {
            samp[i].real = 0.0;
            samp[i].imag = 0.0;
        }

        fft(samp,M);

        for(j = 0 ; j < FFT_LENGTH/2 ; j++) {
            tempflt  = samp[j].real * samp[j].real;
            tempflt += samp[j].imag * samp[j].imag;
            mag[j] = tempflt;
        }
/* find the biggest magnitude spectral bin and output */
        tempflt = mag[0];
        i=0;
        for(j = 1 ; j < FFT_LENGTH/2 ; j++) {
            if(mag[j] > tempflt) {
                tempflt = mag[j];
                i=j;
            }
        }
/* interpolate the peak loacation */
        if(i == 0) {
            p1 = p2 = 0.0;
        }
        else {
            p1 = mag[i] - mag[i-1];
            p2 = mag[i] - mag[i+1];
        }
        sendout(((float)i + 0.5*((p1-p2)/(p1+p2+1e-30)))/FFT_LENGTH);
    }
}
```

LISTING 5.19 (*Continued*)

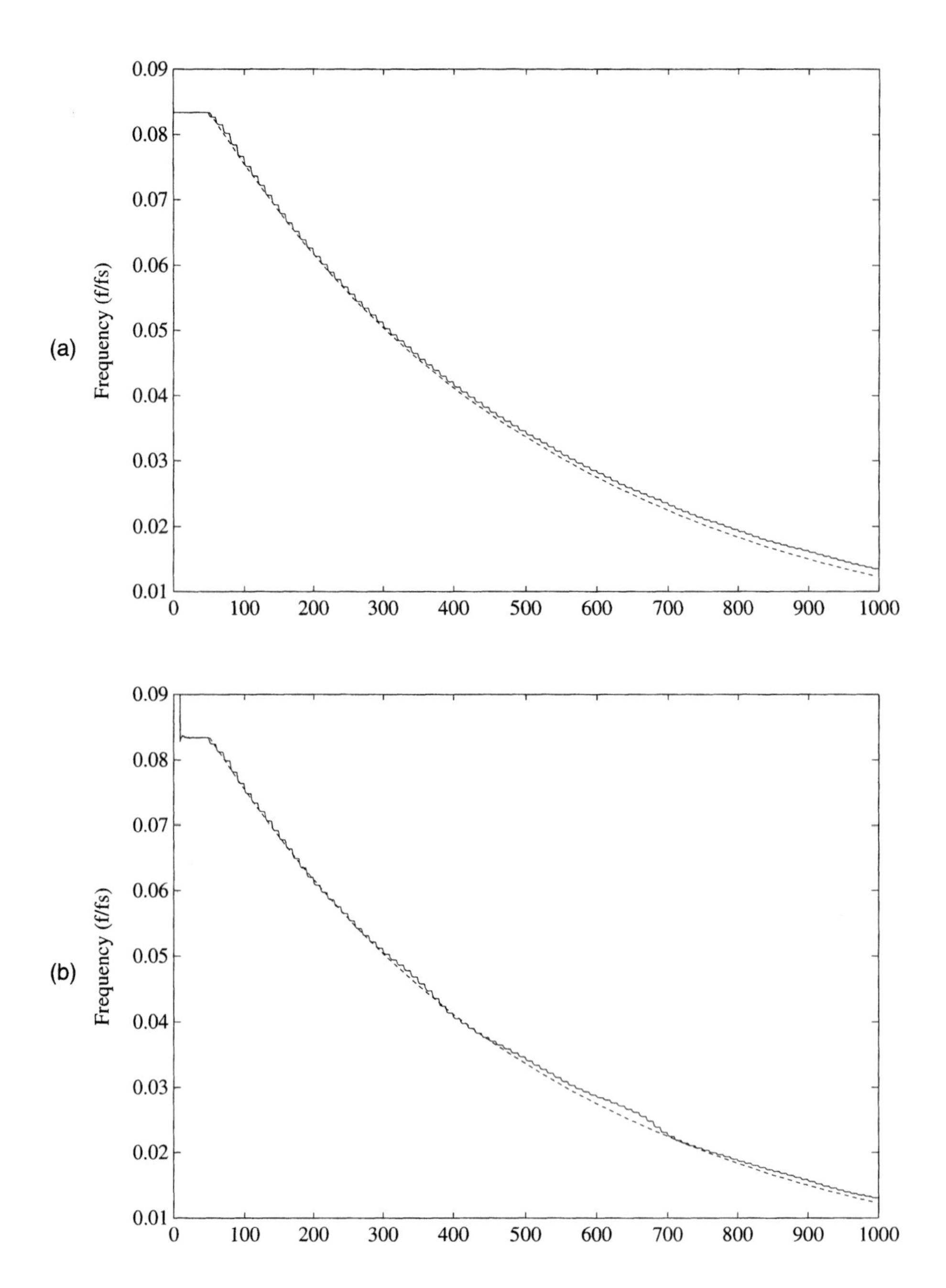

FIGURE 5.15 **(a)** Frequency estimates obtained from program INSTF.C (solid line) with input from OSC.C and correct frequencies (dashed line) generated by OSC.C. The INSTF.C program shown in Listing 5.19 was modified to not add noise to the input signal. **(b)** Frequency estimates obtained from program INSTF.C (solid line) with input from OSC.C and correct frequencies (dashed line) generated by OSC.C. Noise with a standard deviation of 100 was added to the signal from the OSC program.

5.6 REFERENCES

MOORER, J. (August 1977). Signal Processing Aspects of Computer Music: A Survey. *Proceedings of IEEE, 65,* (8).

ALLES, H. (April 1980). Music Synthesis Using Real Time Digital Techniques. *Proceedings of the IEEE, 68,* (4).

SMITH J. and GOSSETT, P. (1984). A Flexible Sample Rate Conversion Method. *Proceedings of ICASSP.*

CROCHIERE, R. and RABINER, L. (March 1981). Interpolation and Decimation of Digital Signals—A Tutorial Review. *Proceedings of the IEEE, 69,* 300–331.

SKOLNIK, M. (1980). *Introduction to Radar Systems,* (2nd ed.). New York: McGraw-Hill.

General Aspects of Digital Transmission Systems (Nov. 1988). Terminal Equipments Recommendations G.700–G.795. International Telegraph and Telephone Consultative Committee (CCITT) 9th Plenary Assembly, Melbourne.

APPENDIX

DSP FUNCTION LIBRARY AND PROGRAMS

The enclosed disk is an IBM-PC compatible high-density disk (1.44 MBytes capacity) and contains four directories called PC, ADSP21K, DSP32C, and C30 for the specific programs that have been compiled and tested for the four platforms discussed in this book. Each directory contains a file called READ.ME, which provides additional information about the software. A short description of the platforms used in testing associated with each directory is as follows:

Directory Name	Platform used to Compile and Test Programs	Available MIPs	Sampling Rate (kHz)
PC	General Purpose IBM-PC or workstation (ANSI C)	Not Real-time	Any
ADSP21K	Analog Devices EZ-LAB ADSP-21020/ADSP-21060 (version 3.1 compiler software)	25	32
DSP32C	CAC AC4-A0 Board with DBDADA-16 (version 1.6.1 compiler software)	12.5	16
C30	Domain Technologies DSPCard-C31 (version 4.50 compiler software)	16.5	16

The following table is a program list of the C programs and functions described in detail in chapters 3, 4, and 5. The first column gives the section number in the text where the program is described and then a short description of the program. The remaining columns give the filenames of the four different versions of the source code for the four different platforms. Note that the files from each platform are in different directories as shown in the previous table.

	PC filename (*.c)	210X0 filename (*.c)	DSP32C filename (*.c)	320C30 filename (*.c)
3.3.3 1024-Point FFT Test Function	fft1k	fftn	fft1k	fft1k
3.4.2 Interrupt-Driven Output example	NA	intout	NA	NA
4.1.1 FIR Filter Function (fir_filter)	filter	filter	filter	filter
4.1.2 FIR Filter Coefficient by Kaiser Window	ksrfir	NA	NA	NA
4.1.2 FIR Filter Coefficients by Parks-McClellan	remez	NA	NA	NA
4.1.3 IIR Filter Function (iir_filter)	filter	filter	filter	filter
4.1.4 Real-Time getinput Function (ASCII text for PC)	getsend	getinput	getinput	send_c30
4.1.4 Real-Time getinput Function (WAV file format)	getwav	NA	NA	NA
4.1.4 Real-Time sendout Function (ASCII text for PC)	getsend	sendout	sendout	send_c30
4.1.4 Real-Time sendout Function (WAV file format)	sendwav	NA	NA	NA
4.2.1 Gaussian Noise Generation Function	filter	filter	filter	filter
4.2.2 Signal-to-Noise Ratio Improvement	mkgwn	mkgwn	mkgwn	mkgwn
4.3.3 Sample Rate Conversion example	interp3	NA	NA	NA
4.4.1 Fast Convolution Using FFT Methods	rfast	rfast21	rfast32	rfast30
4.4.2 Interpolation Using the FFT	intfft2	NA	NA	NA
4.5.1 IIR Filters as Oscillators	osc	osc	osc	osc
4.5.2 Table-Generated Waveforms	wavetab	wavetab	wavetab	wavetab
5.1.1 Speech Spectrum Analysis	rtpse	NA	NA	NA
5.1.2 Doppler Radar Processing	radproc	NA	NA	NA
5.2.1 ARMA Modeling of Signals	arma	NA	NA	NA
5.2.2 AR Frequency Estimation	arfreq	NA	NA	NA
5.3.1 Speech Compression	mulaw	mulaw	mulaw	mulaw
5.3.2 ADPCM (G.722 fixed-point)	g722	g722_21k	NA	g722c3
5.3.2 ADPCM (G.722 floating-point)	NA	g722_21f	g722_32c	g722c3f
5.4.1 Equalization and Noise Removal	equaliz	equaliz	equaliz	equaliz
5.4.2 Pitch-Shifting	pshift	pshift	pshift	pshift
5.4.3 Music Synthesis	music	mu21k	mu32c	muc3
5.5.1 LMS Signal Enhancement	lms	NA	NA	NA
5.5.2 Frequency Tracking with Noise	instf	NA	NA	NA

Note: "NA" refers to programs that are not applicable to a particular hardware platform.

Make files (with an extension .MAK) are also included on the disk for each platform. If the user does not have a make utility availible, PC batch files (with an extension .BAT) are also included with the same name as the make file. The following table is a make file list for many of the C programs described in detail in Chapters 3, 4 and 5:

	PC filename (*.mak)	210X0 filename (*.mak)	DSP32C filename (*.mak)	320C30 filename (*.mak)
3.4.2 Interrupt-Driven Output Example	NA	iout21k	NA	NA
4.2.2 Signal-to-Noise Ratio Improvement	mkgwn	mkgwn	mkgwn	mkgwn
4.3.3 Sample Rate Conversion Example	interp3	NA	NA	NA
4.4.1 Fast Convolution Using FFT Methods	rfast	rf21k	rf32	rfc30
4.4.2 Interpolation Using the FFT	intfft2	NA	NA	NA
4.5.1 IIR Filters as Oscillators	osc	osc21k	osc	osc
4.5.2 Table Generated Waveforms	wavetab	wavetab	wavetab	wavetab
5.1.1 Speech Spectrum Analysis	rtpse	NA	NA	NA
5.1.2 Doppler Radar Processing	radproc	NA	NA	NA
5.2.1 ARMA Modeling of Signals	arma	NA	NA	NA
5.2.2 AR Frequency Estimation	arfreq	arfreq	arfreq	arfreq
5.3.1 Speech Compression	mulaw	mulaw	mulaw	mulaw
5.3.2 ADPCM (G.722 fixed-point)	g722	g722_21k	NA	g722c3
5.3.2 ADPCM (G.722 floating-point)	NA	g722_21f	g722_32c	g722c3f
5.4.1 Equalization and Noise Removal	eqpc	eq	eq	eq
5.4.2 Pitch Shifting	ps	ps	ps	ps
5.4.3 Music Synthesis	music	mu21k	mu32c	muc3
5.5.1 LMS Signal Enhancement	lms	NA	NA	NA
5.5.2 Frequency Tracking with Noise	instf	instf	instf	instf

Note: "NA" refers to programs that are not applicable to a particular platform.

INDEX

G

H

I

K

L

M

N

T

U

V

W

Z

LICENSE AGREEMENT AND LIMITED WARRANTY

READ THE FOLLOWING TERMS AND CONDITIONS CAREFULLY BEFORE OPENING THIS DISK PACKAGE. THIS LEGAL DOCUMENT IS AN AGREEMENT BETWEEN YOU AND PRENTICE-HALL, INC. (THE "COMPANY"). BY OPENING THIS SEALED DISK PACKAGE, YOU ARE AGREEING TO BE BOUND BY THESE TERMS AND CONDITIONS. IF YOU DO NOT AGREE WITH THESE TERMS AND CONDITIONS, DO NOT OPEN THE DISK PACKAGE. PROMPTLY RETURN THE UNOPENED DISK PACKAGE AND ALL ACCOMPANYING ITEMS TO THE PLACE YOU OBTAINED THEM FOR A FULL REFUND OF ANY SUMS YOU HAVE PAID.

1. **GRANT OF LICENSE:** In consideration of your payment of the license fee, which is part of the price you paid for this product, and your agreement to abide by the terms and conditions of this Agreement, the Company grants to you a nonexclusive right to use and display the copy of the enclosed software program (hereinafter the "SOFTWARE") on a single computer (i.e., with a single CPU) at a single location so long as you comply with the terms of this Agreement. The Company reserves all rights not expressly granted to you under this Agreement.

2. **OWNERSHIP OF SOFTWARE:** You own only the magnetic or physical media (the enclosed disks) on which the SOFTWARE is recorded or fixed, but the Company retains all the rights, title, and ownership to the SOFTWARE recorded on the original disk copy(ies) and all subsequent copies of the SOFTWARE, regardless of the form or media on which the original or other copies may exist. This license is not a sale of the original SOFTWARE or any copy to you.

3. **COPY RESTRICTIONS:** This SOFTWARE and the accompanying printed materials and user manual (the "Documentation") are the subject of copyright. You may *not* copy the Documentation or the SOFTWARE, except that you may make a single copy of the SOFTWARE for backup or archival purposes only. You may be held legally responsible for any copying or copyright infringement which is caused or encouraged by your failure to abide by the terms of this restriction.

4. **USE RESTRICTIONS:** You may *not* network the SOFTWARE or otherwise use it on more than one computer or computer terminal at the same time. You may physically transfer the SOFTWARE from one computer to another provided that the SOFTWARE is used on only one computer at a time. You may *not* distribute copies of the SOFTWARE or Documentation to others. You may *not* reverse engineer, disassemble, decompile, modify, adapt, translate, or create derivative works based on the SOFTWARE or the Documentation without the prior written consent of the Company.

5. **TRANSFER RESTRICTIONS:** The enclosed SOFTWARE is licensed only to you and may *not* be transferred to any one else without the prior written consent of the Company. Any unauthorized transfer of the SOFTWARE shall result in the immediate termination of this Agreement.

6. **TERMINATION:** This license is effective until terminated. This license will terminate automatically without notice from the Company and become null and void if you fail to comply with any provisions or limitations of this license. Upon termination, you shall destroy the Documentation and all copies of the SOFTWARE. All provisions of this Agreement as to warranties, limitation of liability, remedies or damages, and our ownership rights shall survive termination.

7. **MISCELLANEOUS:** This Agreement shall be construed in accordance with the laws of the United States of America and the State of New York and shall benefit the Company, its affiliates, and assignees.

8. **LIMITED WARRANTY AND DISCLAIMER OF WARRANTY:** The Company warrants that the SOFTWARE, when properly used in accordance with the Documentation, will operate in substantial conformity with the description of the SOFTWARE set forth in the Documentation. The Company does not warrant that the SOFTWARE will meet your requirements or that the operation of the SOFTWARE will be uninterrupted or error-free. The Company warrants that the media on which the SOFTWARE is delivered shall be free from defects in materials and workmanship under normal use for a period of thirty (30) days from the date of your purchase. Your only remedy and the Company's only obligation under these limited warranties is, at the Company's option, return of the warranted item for a refund of any amounts paid by you or replacement of the item. Any replacement of SOFTWARE or media under the warranties shall not extend the original warranty pe-

riod. The limited warranty set forth above shall not apply to any SOFTWARE which the Company determines in good faith has been subject to misuse, neglect, improper installation, repair, alteration, or damage by you. EXCEPT FOR THE EXPRESSED WARRANTIES SET FORTH ABOVE, THE COMPANY DISCLAIMS ALL WARRANTIES, EXPRESS OR IMPLIED, INCLUDING WITHOUT LIMITATION, THE IMPLIED WARRANTIES OF MERCHANTABILITY AND FITNESS FOR A PARTICULAR PURPOSE. EXCEPT FOR THE EXPRESS WARRANTY SET FORTH ABOVE, THE COMPANY DOES NOT WARRANT, GUARANTEE, OR MAKE ANY REPRESENTATION REGARDING THE USE OR THE RESULTS OF THE USE OF THE SOFTWARE IN TERMS OF ITS CORRECTNESS, ACCURACY, RELIABILITY, CURRENTNESS, OR OTHERWISE.

IN NO EVENT, SHALL THE COMPANY OR ITS EMPLOYEES, AGENTS, SUPPLIERS, OR CONTRACTORS BE LIABLE FOR ANY INCIDENTAL, INDIRECT, SPECIAL, OR CONSEQUENTIAL DAMAGES ARISING OUT OF OR IN CONNECTION WITH THE LICENSE GRANTED UNDER THIS AGREEMENT, OR FOR LOSS OF USE, LOSS OF DATA, LOSS OF INCOME OR PROFIT, OR OTHER LOSSES, SUSTAINED AS A RESULT OF INJURY TO ANY PERSON, OR LOSS OF OR DAMAGE TO PROPERTY, OR CLAIMS OF THIRD PARTIES, EVEN IF THE COMPANY OR AN AUTHORIZED REPRESENTATIVE OF THE COMPANY HAS BEEN ADVISED OF THE POSSIBILITY OF SUCH DAMAGES. IN NO EVENT SHALL LIABILITY OF THE COMPANY FOR DAMAGES WITH RESPECT TO THE SOFTWARE EXCEED THE AMOUNTS ACTUALLY PAID BY YOU, IF ANY, FOR THE SOFTWARE.

SOME JURISDICTIONS DO NOT ALLOW THE LIMITATION OF IMPLIED WARRANTIES OR LIABILITY FOR INCIDENTAL, INDIRECT, SPECIAL, OR CONSEQUENTIAL DAMAGES, SO THE ABOVE LIMITATIONS MAY NOT ALWAYS APPLY. THE WARRANTIES IN THIS AGREEMENT GIVE YOU SPECIFIC LEGAL RIGHTS AND YOU MAY ALSO HAVE OTHER RIGHTS WHICH VARY IN ACCORDANCE WITH LOCAL LAW.

ACKNOWLEDGMENT

YOU ACKNOWLEDGE THAT YOU HAVE READ THIS AGREEMENT, UNDERSTAND IT ,AND AGREE TO BE BOUND BY ITS TERMS AND CONDITIONS. YOU ALSO AGREE THAT THIS AGREEMENT IS THE COMPLETE AND EXCLUSIVE STATEMENT OF THE AGREEMENT BETWEEN YOU AND THE COMPANY AND SUPERSEDES ALL PROPOSALS OR PRIOR AGREEMENTS, ORAL, OR WRITTEN, AND ANY OTHER COMMUNICATIONS BETWEEN YOU AND THE COMPANY OR ANY REPRESENTATIVE OF THE COMPANY RELATING TO THE SUBJECT MATTER OF THIS AGREEMENT.

Should you have any questions concerning this Agreement or if you wish to contact the Company for any reason, please contact in writing at the address below or call the at the telephone number provided.

PTR Customer Service
Prentice Hall PTR
One Lake Street
Upper Saddle River, New Jersey 07458

Telephone: 201-236-7105